Electrical Solitons

Theory, Design, and Applications

Devices, Circuits, and Systems

Series Editor
Krzysztof Iniewski
CMOS Emerging Technologies Inc., Vancouver, British Columbia, Canada

Internet Networks: Wired, Wireless, and Optical Technologies
Krzysztof Iniewski

Semiconductor Radiation Detection Systems
Krzysztof Iniewski

Electronics for Radiation Detection
Krzysztof Iniewski

Radiation Effects in Semiconductors
Krzysztof Iniewski

Electrical Solitons: Theory, Design, and Applications
David Ricketts and Donhee Ham

FORTHCOMING

Radio Frequency Integrated Circuit Design
Sebastian Magierowski

Semiconductors: Integrated Circuit Design for Manufacturability
Artur Balasinki

Integrated Microsystems: Materials, MEMs, Photonics, Bio Interfaces
Krzysztof Iniewski

Electrical Solitons

Theory, Design, and Applications

David S. Ricketts
Donhee Ham

CRC Press
Taylor & Francis Group
Boca Raton London New York

CRC Press is an imprint of the
Taylor & Francis Group, an **informa** business

CRC Press
Taylor & Francis Group
6000 Broken Sound Parkway NW, Suite 300
Boca Raton, FL 33487-2742

Printed in the United States of America on acid-free paper
10 9 8 7 6 5 4 3 2 1

International Standard Book Number: 978-1-4398-2980-6 (Hardback)

Library of Congress Cataloging-in-Publication Data

Ricketts, David Shawn
Electrical solitons : theory, design, and applications / David S. Ricketts, Donhee Ham.
p. cm. -- (Devices, circuits, and systems)
Includes bibliographical references and index.
ISBN 978-1-4398-2980-6 (hardback)
1. Solitons. I. Ham, Donhee. II. Title.

QC174.26.W28R53 2010
530.12'4--dc22 2010042045

Visit the Taylor & Francis Web site at
http://www.taylorandfrancis.com

and the CRC Press Web site at
http://www.crcpress.com

Contents

List of Figures

List of Tables

Preface

This book developed out of the combined research of our groups at Carnegie Mellon University and Harvard University. While many texts have been written about solitons in general, and about KdV solitons, no text exists today that focuses on a comprehensive treatment of the soliton in the electrical domain, namely the KdV soliton on the nonlinear transmission line (NLTL). The NLTL is the dominant medium for soliton propagation in electronics and has found wide application as a testbed for nonlinear dynamics and KdV phenomena as well as for practical applications in ultra-sharp pulse/edge generation and novel nonlinear communication schemes in electronics.

Solitons exist in many other media, such as water, optical fibers and spring-mass systems. Although there are other solitons in the "electrical" domain, namely magnetostatic spin-waves in magnetic thin films and fluxons in Josephson junctions, these are not based upon KdV dynamics, but rather the nonlinear Schrödinger equation and the Sine-Gordon equation, respectively. They have different characteristics than KdV solitons, although they share the same foundational mathematical structure described by the inverse scattering method. Our discussion of "Electrical Solitons" in this book as the KdV soliton on the NLTL is therefore not comprehensive, but in terms of voltage waves, it is certainly the dominant form of soliton in the electrical domain.

This book describes, in a comprehensive manner, the theory, design and application of electrical solitons and is composed of three parts.

Part I: This part describes the foundational theory for KdV solitons, which is the basis for voltage solitons on the NLTL. Chapter 2 describes the solution to the KdV equation and the basic properties of that solution, including collision behaviors and amplitude-dependent velocity. Chapter 3 presents the core underlying mathematics of solitons in an easy-to-read presentation. These chapters represent a distillation of a large number of rather lengthy and mathematical sources. We have strived to provide an understandable explanation for the reader interested in electrical KdV solitons. This part ends with Chapter 4, which discusses the conservation laws of the KdV for loss-less and lossy systems.

Part II: This part describes the KdV soliton in the context of the NLTL. Chapter 5 derives the lattice equation for solitons on the NLTL and shows the connection with the KdV equation. It also derives the governing equations for a lossy NLTL. Chapter 6 translates the theory into practical implementation in the laboratory. It details the transformation between what we measure on the oscilloscope and KdV theory. It shows explicit demonstrations of many of the key properties of solitons, including the inverse scattering method and soliton damping. It provides a practical demonstration of the theory developed in Part I.

Part III: This part contains six chapters on the application of the electrical soliton. The first chapter, Chapter 7, describes several applications of electrical solitons and the NLTL in sharp pulse formation and edge sharpening for high speed metrology as well as high frequency generation via NLTL harmonics. Chapter 8 describes the challenges to realizing a robust soliton oscillator and the stability mechanisms necessary for its development. Chapter 9 introduces three prototypes of the circular soliton oscillator using discrete and integrated platforms. These works demonstrate the unique dynamics and broad implementation of the concept of the soliton oscillator. Chapter 10 describes a reflection-based soliton oscillator that uses the same stabilizing techniques as the circular soliton oscillator, but extends them to a new NLTL architecture. Chapter 11 describes an adaptation of the soliton oscillator for a chaotic communication. In these systems the stability mechanisms are removed and the oscillator is allowed to become chaotic, which is then leveraged for secure communication. We conclude our discussion of soliton oscillators with Chapter 12, which details the theory of phase noise of these new oscillating systems.

Finally, this book and the work it represents would not be possible without all of our supporters, both professional and personal. We wish to thank our parents and teachers throughout our education. We have included many of the students who contributed to these works as chapter co-authors of their respective works; however, there are many, many more whom we have worked with and gained insight and support from, and we wish to thank them all.

This book has been a joy to prepare and we hope the reader will appreciate the beauty of the soliton and enjoy what we hope to be a comprehensive discussion of its application in electronics.

Pittsburgh, PA — *David S. Ricketts*
Cambridge, MA — *Donhee Ham*

Part I

Electrical Solitons: Theory

1

Introduction

CONTENTS

Solitons are a special class of pulse-shaped waves that propagate without changing their shape in nonlinear dispersive media [1–5]. The term soliton was coined by Zabusky and Kruskal in their 1965 paper on the collision of waves in plasma [2] where they observed a particle-like property of the wave phenomenon. A balancing mechanism between nonlinearity and dispersion is responsible for the soliton phenomena. Solitons take on a sech^2 pulse shape and propagate in a nonlinear dispersive medium without changing shape. Their velocities are parameter-dependent, such as amplitude-dependent or color (frequency) dependent, and as a result, solitons of differing parameters can collide with one another. One would typically expect the generation or new waveforms/content when two waveforms are mixed, i.e., collided, in a nonlinear medium. Interestingly, when solitons collide with other solitons they do experience a nonlinear collision but re-emerge fully intact as the exact solitons that entered the collision. This is counterintuitive as we often expect a nonlinear interaction to permanently modify the waveform, e.g., create harmonics, etc. The soliton collision is reminiscent of two objects colliding, such as tennis balls, which may deform upon collision, but emerge fully intact and unaltered [4]. This interpretation yields the "particle" viewpoint of these solitary waves, i.e., solitons. An example of such a collision is shown in Fig. 1.1. The ability of solitons to undergo nonlinear interaction and re-emerge undisturbed provides a glimpse of the underlying beauty of their dynamics. In Chapter 3 we will explore in detail the deep, underlying mathematics of solitons that is responsible for their unique behavior.

Nature offers a variety of soliton examples. The first-reported soliton was a solitary water wave in a narrow canal where the shallow water possessed both nonlinearity and dispersion [5, 6]. A mechanical spring-mass lattice can also act as a nonlinear dispersive medium, propagating solitons in the form of lattice waves [7,8]. The optical fiber is yet another

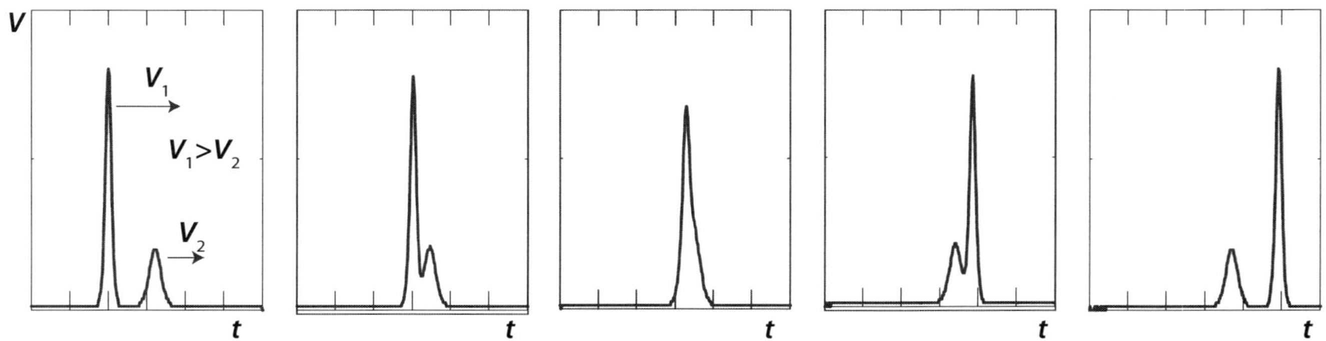

FIGURE 1.1
Time series of two KdV solitons colliding. Due to the amplitude-dependent velocity of KdV solitons, a taller soliton will overtake a shorter soliton and in the process collide with it. During the nonlinear collision the waveform "distorts," i.e., is not a linear sum of the two waveforms, but as the taller soliton emerges from the collision, the waveform returns to its original form, i.e., two exact solitons (except for a phase shift in position).

example of a nonlinear dispersive medium where optical solitons are observed [9].

In electronics, the nonlinear transmission line (NLTL) serves as a nonlinear dispersive medium that propagates voltage solitons [10, 11] (See Chapter 5). Electrical solitons on the NLTL have been actively investigated over the last 40 years, particularly in the microwave domain, for sharp pulse generation applications [11, 12] and for high-speed RF and microwave sampling applications [11, 13, 14]. In ultra-fast time-domain metrology, the short duration of a soliton directly translates to high temporal resolution. For example, narrow electrical pulses can be used to sample, or take "snapshots" of, rapidly varying electrical signals [11,13], as shown in Fig. 1.2(a). For another example, the sharp, narrow electrical pulses can be used as probe signals for high-precision time-domain reflectometry (TDR) [15], Fig. 1.2(b). In addition to high-resolution metrology, periodic sharp pulse trains can also be used in high-speed UWB and impulse, radar and ranging systems. More recently, new soliton systems have been developed that self-generate periodic soliton pulses [16], as well as several novel communication schemes based upon the electrical solitons [17, 18].

In the electrical domain, the electrical soliton, which propagates on the NLTL, is governed by the dynamics of the KdV equation[1]. This book develops the theory, design and application of the KdV soliton on the NLTL. The KdV soliton can be found in many other areas such as plasma, hydrodynamics, and mechanical spring-mass systems and many of the concepts developed in this book in the context of NLTL KdV solitons are applicable. The next section overviews the three canonical solitons to place the KdV soliton in perspective.

[1]More precisely the NLTL is a Toda lattice, which is an exact soliton system that reduces to the KdV system for small amplitudes. Since the KdV system is a good approximation, yields all of the correct physical insights and has a much more developed body of theory, we choose to use it as the basis for electrical solitons.

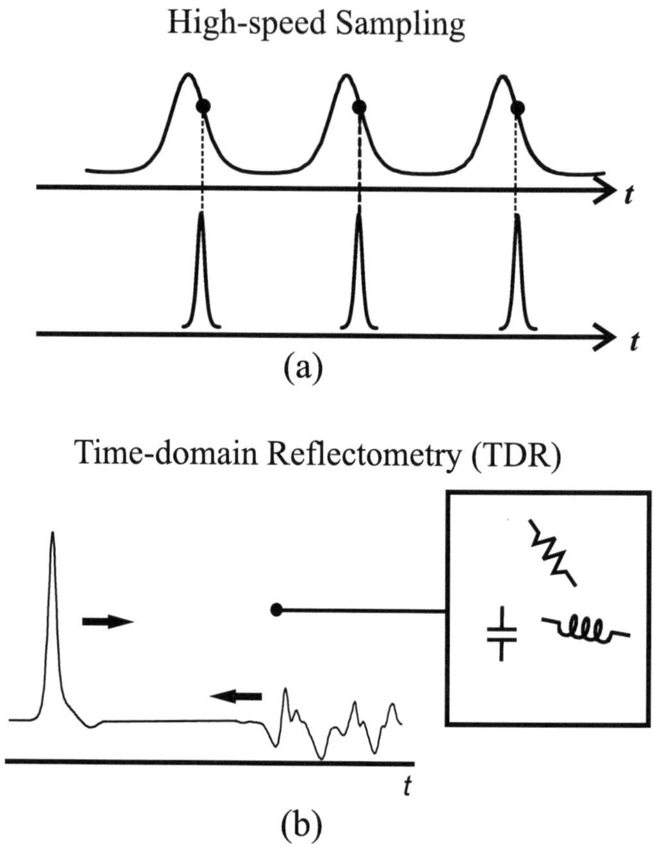

FIGURE 1.2
(a) Electrical pulses can be used to sample rapidly varying electrical signals. (b) The narrow width and wide bandwidth of the electrical pulses allow for the measurement of both the composition and the distribution of complex networks.

1.1 The "Solitons"

The soliton phenomenon has been found in many other systems in nature. Most of these soliton systems are based on the dynamics of one of three "classical" soliton equations [19]:

- **Korteweg-de Vries** equation (**KdV**). This equation models the dynamics of shallow water waves, nonlinear spring-mass lattices, plasma, and electrical solitons;

$$u_t + 6uu_x + u_{xxx} = 0.$$

- **Nonlinear Schrödinger** equation (**NLS**). This equation models the dynamics of spin-waves in magnetic materials and light in fiber-optic cables;

$$iA_x - A_{tt} + 2|A|^2 A = 0.$$

- **Sine-Gordon** equation (**SG**). This equation models the dynamics of Josephson junctions and nonlinear pendulum lattices;

$$\phi_{tt} - \phi_{xx} + \sin\phi = 0.$$

Figure 1.3 shows the solutions to these three equations: the KdV soliton pulse, the NLS envelope soliton, and the SG kink soliton.

1.2 A Brief Overview and History of the Soliton

At the heart of this work are the fascinating dynamics of the soliton phenomenon. In this section we give a brief history of the soliton from its discovery to its modern implementation in electronics. Throughout the book we will refer back to these works as they relate to the specific topic being discussed.

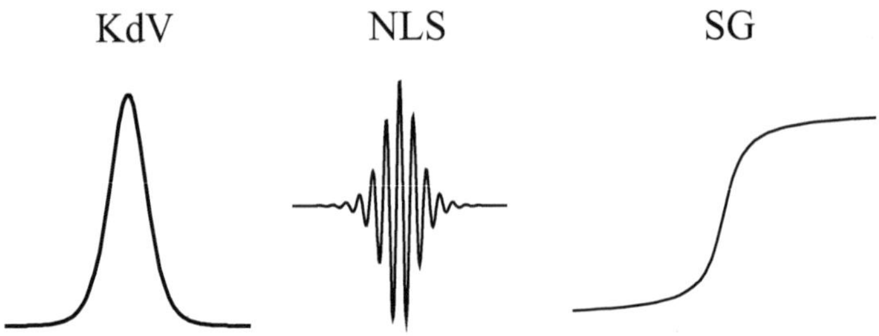

FIGURE 1.3
Classical solitons: KdV soliton, NLS envelope soliton, and SG kink soliton.

1834 (*Hydrodynamics*) John Scott Russell was towing a barge with a pair of horses through a narrow channel when the barge suddenly stopped. The bow wave, however, continued and Russell followed it on horseback for several miles, noticing that the wave maintained its single pulse state while propagating for large distances. This is the first noted observation of a soliton, which Russell named "The Wave of Translation." His observations and further analysis of this wave were published in [6].

1895 (*Hydrodynamics*) Korteweg and de Vries wrote their famous paper on water wave dynamics [20]. In it they developed what is known today as the Korteweg-de Vries (KdV) equation for the evolution of shallow water waves. One of the key contributions of this paper was that it provided an analytical solution for Russell's "Wave of Translation" based upon the physics of water waves. Korteweg and de Vries refer to this single wave solution as a solitary wave.

1955 (*Anharmonic lattices*) Enrico Fermi and colleagues, J. Pasta and S. Ulam, were interested in examining the distribution of energy in crystal lattices, in particular the redistribution of energy from one initial mode to equal energies in all modes, i.e., energy equipartition [7]. In particular they wanted to test a theory proposed by Debye in 1914, that the nonlinear potential between atoms would cause energy in a single mode to distribute over all modes after a transient period [5]. To test this theory, they constructed a 1-D spring-mass lattice with 64 masses, where the spring constant was nonlinear, and simulated the lattice using a new device known as the computer. Initially everything looked good as they saw an initial sinusoidal mode start to break apart in what appeared to be multiple modes, i.e., redistribution of the energy across multiple modes. The story goes that they all went for a break and left the simulator running [19]; when they returned they saw that the original sinusoidal shape had reappeared. This phenomena became known as the "reoccurrence phenomena" and is found in many soliton systems, including electrical solitons [21]. Later it was learned that this phenomena is caused by the initial mode breaking up into multiple solitons that travel at different speeds. These solitons eventually collide with one another (the taller soliton catches up with the shortest soliton) and the original profile is recreated. This was the first glimpse of some of the deeper properties of solitons. The electrical lattice we will use to propagate electrical solitons is an electrical analog to the one studied by Fermi and his colleagues.

1955 (*Electrical*) At the same time that Fermi was investigating his lattice, C. C. Cutler was developing a new field that would become known as *modelocking*. In Cutler's work he developed the first regenerative pulse generator that produced a periodic train of electrical pulses [22], i.e., a modelocked oscillator. In his work, he combined a linear transmission line, i.e., *non*-soliton, with an amplifier and a novel construct that was able to sharpen linear electrical pulses. This sharpening mechanism was achieved through a threshold-dependent gain-attenuation mechanism placed in the oscillator loop. This mechanism is known today as *saturable absorbtion* and is used extensively in optical modelocking. The work presented in Chapters 8–10 uses the same mechanism, but for a different purpose than Cutler's — the stabilization of solitons in a closed-loop oscillator.

1960 (*Electrical*) Rolf Landauer introduced an electrical structure known as the nonlinear transmission line for parametric amplification [23]. It was later shown by several scientists that this nonlinear transmission line was a direct analog of Fermi's nonlinear lattice and, as such, could propagate solitons in the form of voltage waves — these are the solitons that we will study in this book.

1965 (*Plasma physics*) Zabusky and Kruskal wrote their famous paper on the interaction of solitary waves in plasma. Plasma solitons are well modeled by the KdV equation and are very similar to the solitons in Fermi's nonlinear spring-mass lattice [2] and the electrical solitons we will study in this book. It is in Zabusky's and Kruskal's paper that they coined the term "soliton," which was meant to bring together the ideas of a *solitary wave* and a *particle*, such as an electron or proton.

1966 (*Optics*) DeMaria and colleagues adapted Cutler's technique to the optical domain to create the first optical self-modelocking system [24]. In particular, DeMaria used the concept of *saturable absorption* for the generation of the optical pulses. Optical *soliton* modelocking did not come about until 1992 when it was introduced by Tamura and colleagues [25]. The soliton oscillator discussed in this book mirrors the development of soliton modelocking in optics.

1967 (*Anharmonic lattice*) M. Toda introduced a spring-mass lattice similar to Fermi's, except that the force between masses follows an exponential relationship [26, 27]. This exponential relationship is *key*, as Toda shows that an **exact** soliton system is created. The Toda lattice, as it became known, supports soliton propagation and exhibits all of the dynamics of soliton systems. In the small amplitude and long-wavelength limit, the Toda lattice reduces to the KdV equation. The electrical soliton we will use in this book is an approximation between Toda lattice solitons and KdV solitons.

1967 (*Mathematics*) Gardner, Greene, Kruskal and Miura introduced an amazing mathematical method for solving the initial value problem of the KdV equation, which is known as the Inverse Scattering Method [28] (Korteweg and de Vries found the steady state solution in 1895). This method not only allowed for the solution of an initial value problem, but also introduced the concept that each soliton could be represented by a time-invariant eigenvalue. These time-invariant eigenvalues are what give solitons their unique properties, in fact in Chapter 3 we will define the soliton through these eigenvalues.

1968 (*Mathematics*) Lax developed his formalism for the Inverse Scattering Method [29]. This formalism provided a comprehensive method for explaining the inverse scattering method and allowed others to extend the inverse scattering method to other soliton equations, including the nonlinear Schrödinger and the sine-Gordon equations.

1970s and '80s (*Electronics*) Many researchers investigated the dynamics of solitons by using the nonlinear transmission line. At first, this was done as an analog to Toda's mechanical lattice, since a discrete L-C network with nonlinear capacitors was an excellent experimental model for the Toda lattice (See Hirota [10]). Researchers soon learned, however, that this structure and the solitons it propagates, could be used for many engineering applications due to the narrow width of the electrical solitons produced (See Jäger [30]).

1973 (*Optics*) Hasegawa proposed adding nonlinearity to dispersive optical fibers [9] which eventually lead to the development of optical solitons in 1980 [31]. It wasn't until 1992, however, that optical solitons where used in mode-locking [25]. Optical solitons are based on the nonlinear Schrödinger equation.

1988 (*Electronics*) Tan introduced a tapered nonlinear transmission line, where the dispersion of the line was gradually reduced along the line. This line was able to compress a input pulse by 7× [32]. A non-tapered nonlinear transmission line, in contrast, provides approximately 2× pulse compression. This work paved the way for all-electrical picosecond pulse generation.

1990s (*Electronics*) Mark Rodwell and Daniel van der Weide furthered Tan's technique and integrated it on a monolithic substrate with state-of-the-art GaAs technologies to produce the fastest electrical pulses ever generated from an all-electrical system. Rodwell's student, Case, achieved a pulse width of 5 ps [33] while van der Weide showed an edge rise time of 480 fs [34]. These works still represent the fastest waveforms generated with the nonlinear transmission line. These ultra-fast NLTLs have been used in numerous applications, in particular ultra-fast metrology, where sampling bandwidths of over 700 GHz have been achieved [35]. At the same time as the high-speed applications of solitons were being developed, several other soliton systems were investigated. Singer and Oppenheim at MIT developed a novel soliton based communication scheme for signal compression [17]. Also Ballantyne, Gough and Taylor at the University of Canterbury, New Zealand, investigated soliton dynamics on a closed loop NLTL [36] where a periodic train of solitons was self generated. This system was a precursor to the soliton oscillator developed in [16]; however, it did not contain the necessary stability mechanisms to achieve a robust, stable oscillation.

2000s (*Integrated Electronics*) The past decade brought about many efforts in soliton systems and their implementation in integrated circuits. Ricketts at Carnegie Mellon University and Ham at Harvard University developed the necessary stabilizing mechanisms to construct a robust soliton oscillator, [16,37] and demonstrated this in an integrated prototype, [38]. Work continues today in their research groups on integrated soliton systems, including applications to chaotic systems, reflection oscillators and many more. In addition to the system level work of Ricketts and Ham, Hajimiri at the California Institute of Technology and Afshari at Cornell University extended the 2-port work of Rodwell and van der Weide into integrated circuits with the development of on-chip pulse sharpening [39] and pulse conditioning circuits [40]. They and many other researches continue today with their work on nonlinear circuits and solitons in electronics.

2

The KdV Soliton

CONTENTS

This chapter provides a practical treatment of the theory of KdV solitons and includes the key derivations and references to the literature. The material presented in this chapter is taken from multiple sources: [3, 5, 29, 41, 42], some of which themselves are distillations of larger bodies of previous work. Its presentation here represents a distillation of the key derivations and concepts from these sources as they apply to the electrical soliton. In this sense, it provides a condensed and relevant summary of a much larger body of work. In several instances the methods have been simplified or modified to give a clearer understanding of the relevant materials, and thus may deviate from those methods described in the original references.

2.1 The Solitary Wave Solution

In 1895 Korteweg and de-Vries derived their famous equation for the propagation of waves in shallow water. This equation was significant since it showed in very simple terms the complex balance between nonlinearity and dispersion in shallow water:

$$u_t + 6uu_x + u_{xxx} = 0. \tag{2.1}$$

where we adopt the notation, $\frac{\partial u}{\partial t} = u_t$ and $\frac{\partial u}{\partial x} = u_x$. Moreover, this equation provided for the simple solitary wave solution that matched the experimental observations of Russell's solitary wave:

$$U(x,t) = A \cdot \text{sech}^2 \left[\sqrt{\frac{A}{2}}(x - 2At) \right]. \tag{2.2}$$

As a starting point, we show that (2.2) is a solution to (2.1). Using (2.2) as an ansatz we can see directly the balance between dispersion and nonlinearity in the KdV equation by noting the following identities $\left(\theta = \sqrt{\frac{A}{2}}\,(x - 2At)\right)$:

$$\begin{aligned}
U_t &= -\sqrt{2}A^{3/2} \cdot \text{sech}^2(\theta) \cdot \tanh(\theta) && (2.3)\\
6UU_x &= -6\sqrt{2}A^{5/2} \cdot \text{sech}^4(\theta) \cdot \tanh(\theta) && (2.4)\\
U_{xxx} &= 6\sqrt{2}A^{5/2} \cdot \text{sech}^2(\theta) \cdot \tanh(\theta) \\
&\quad -2\sqrt{2}A^{5/2} \cdot \text{sech}^4(\theta) \cdot \tanh(\theta) \\
&= -2A \cdot U_x - 6UU_x. && (2.5)
\end{aligned}$$

Now with (2.5) we can rewrite the KdV equation, (2.1), for a $U(x,t) = A \cdot \text{sech}^2(\theta)$ waveform as follows:

$$U_t + 6UU_x - 2A \cdot U_x - 6UU_x = 0, \tag{2.6}$$

where the $6UU_x$ term from the dispersion, (2.5), cancels out the nonlinearity to give

$$U_t - 2A \cdot U_x = 0. \tag{2.7}$$

From this we can clearly see the cancelation of nonlinearity by dispersion and the resulting $U(x,t) = A \cdot \text{sech}^2(\theta)$ traveling wave solutions with velocity $2A$.

2.2 The Periodic Soliton: The Cnoidal Wave Solution

In this section we solve the KdV equation for periodic solutions and show that the solitary wave discussed in Sec. 2.1 is simply the limit of the periodic solution as the period goes to infinity.

The derivation of the periodic solution to the KdV equation utilizes elliptic functions, which are unfamiliar to most scientists and engineers. They represent solutions to elliptic integrals which appear in the derivations of many physical systems [43] which have elliptic geometry. Moreover, elliptic integrals provide a means to solve nonlinear, polynomial differential equations. Two prime examples are the large angle solution for the pendulum [44] and also the solution to the KdV equation. Appendix 2.A provides more details and insights into elliptic integrals and functions.

We begin with the KdV equation, where we have rewritten (2.1) for convenience:

$$u_t + 6uu_x + u_{xxx} = 0. \tag{2.8}$$

We assume a traveling wave solution in the reference frame $\xi = x - ct$. This reduces (2.8) to a single variable,

$$-c \cdot u_\xi + 6uu_\xi + u_{\xi\xi\xi} = 0, \tag{2.9}$$

which we can now integrate with respect to u_ξ to give:

$$-c \cdot u + 3u^2 + u_{\xi\xi} = C_1. \tag{2.10}$$

Now by multiplying both sides by u_ξ we can write this in an easily integrable form:

$$-c \cdot uu_\xi + 3u^2 u_\xi + \frac{1}{2}\left(u_\xi\right)_\xi^2 = C_1 u_\xi. \tag{2.11}$$

Integrating this once again gives:

$$-\frac{c}{2}u^2 + u^3 + \frac{1}{2}(u_\xi)^2 = C_1 u + C_2 \tag{2.12}$$

or

$$\begin{aligned} u_\xi &= \sqrt{-2u^3 + cu^2 + 2C_1 u + 2C_2} \\ &= \sqrt{2}\sqrt{-(u-z_1)(u-z_2)(u-z_3)}, \end{aligned} \tag{2.13}$$

$$\tag{2.14}$$

with

$$\begin{aligned} c &= 2(z_1 + z_2 + z_3) && (2.15) \\ C_1 &= -(z_1 z_2 + z_1 z_3 + z_2 z_3) && (2.16) \\ C_2 &= z_1 z_2 z_3 && (2.17) \\ z_1 < z_2 &< z_3. && (2.18) \end{aligned}$$

This can then be integrated assuming periodic boundary conditions,

$$\int_{u_o}^{u} \frac{du}{\sqrt{2}\sqrt{-(u-z_1)(u-z_2)(u-z_3)}} = \int_{\xi_o}^{\xi} d\xi, \tag{2.19}$$

to give the final solution of $u(x,t)$ as (see [5])

$$\begin{aligned} u(x,t) &= z_2 + (z_3 - z_2)cn^2\left[\sqrt{\frac{z_3 - z_1}{2}}\xi; m\right] && (2.20) \\ m &= \frac{z_3 - z_2}{z_3 - z_1}. && (2.21) \end{aligned}$$

This periodic waveform is known as a cnoidal wave. Its amplitude, pulse width and velocity are written in terms of the zeros of the cubic function in (2.13). The solution, (2.20), can be further simplified by assuming a zero offset ($z_2 = 0$) as follows:

$$u(x,t) = z_3 \cdot cn^2\left[\sqrt{\frac{z_3 - z_1}{2}}\xi; m\right]. \tag{2.22}$$

Replacing ξ with $x - ct$ and using (2.15) and (2.21) we get:

$$u(x,t) = z_3 \cdot cn^2\left[\sqrt{\frac{z_3 - z_1}{2}}(x - 2(z_1 + z_3)t); \frac{z_3}{z_3 - z_1}\right], \tag{2.23}$$
$$\tag{2.24}$$

with the full-width-half-maximum (FWHM) pulse width and velocity given as

$$w = 1.76\sqrt{\frac{2}{z_3 - z_1}} \tag{2.25}$$
$$c = 2(z_1 + z_3). \tag{2.26}$$

The cnoidal function is periodic like sinusoidal functions, except instead of being periodic in 2π the cnoidal function is periodic in $2K(m)$, where $K(m)$ is known as the complete elliptic integral of the first kind:

$$cn^2\left[\sqrt{\frac{z_3 - z_1}{2}}\xi; m\right] = cn^2\left[\sqrt{\frac{z_3 - z_1}{2}}\xi + 2K(m)\cdot n; m\right] \tag{2.27}$$
$$n = 0, 1, 2, \ldots$$
$$K(m) = \int_0^{\pi/2} \frac{d\phi}{\sqrt{(1 - m \cdot \sin^2\phi)}}. \tag{2.28}$$

Note that (2.27) allows for multiple solutions, or modes, depending on the value of n. This is like a violin string. The two ends, or boundary conditions, are fixed, however, the string may vibrate in any of the modes which match those boundary conditions, allowing one, two, three or more peaks between the two boundaries. In Chapter 8 we will see the modes when we discuss the oscillator topology.

The wavelength of a cnoidal wave is given as follows:

$$\lambda = 2K\left(\frac{z_3}{z_3 - z_1}\right) \cdot \sqrt{\frac{2}{z_3 - z_1}}. \tag{2.29}$$

The cnoidal wave solution, therefore, is dependent on **two** parameters:

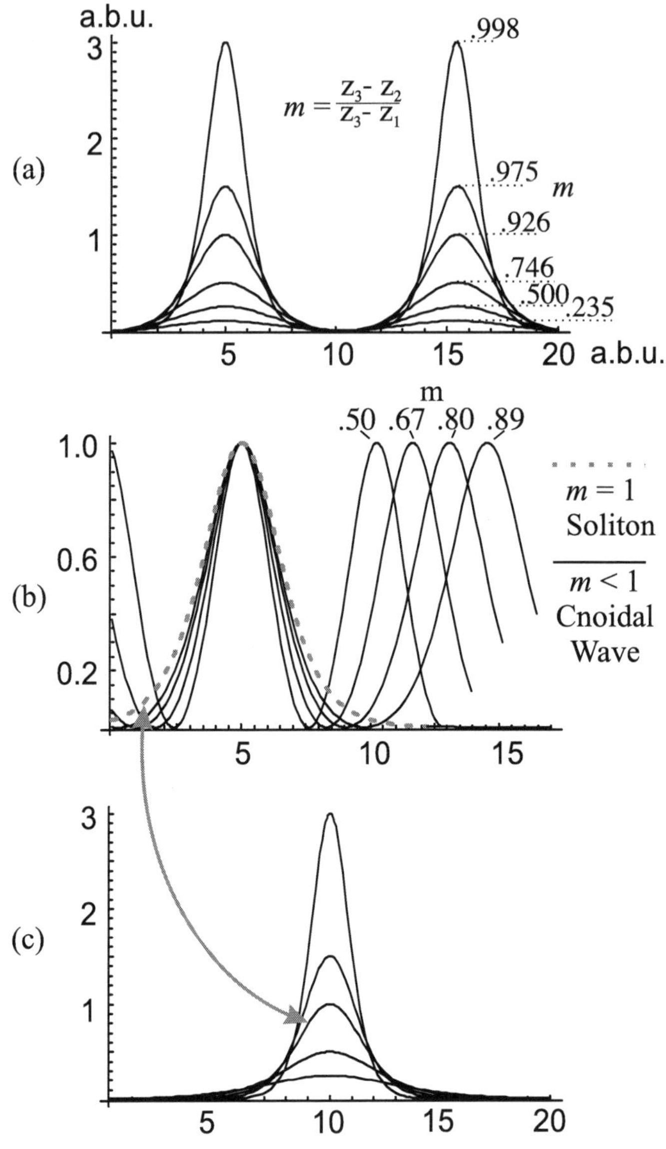

FIGURE 2.1
(a) Shape variation versus amplitude for fixed wavelength, λ. (b) Shape variation versus wavelength for fixed amplitude; note pulse width is wavelength-dependent. The gray dashed line shows the profile of the soliton with an infinite wavelength. Note that the cnoidal wave approaches this shape as the wavelength goes to infinity ($m \to 1$).(c) Amplitude-dependent width of the soliton.

z_1 and z_3, where the first is determined by the the wavelength, λ, and the second by the amplitude. The result is a wave shape and speed that is dependent on both amplitude and wavelength. Figure 2.1 illustrates this somewhat unfamiliar dependence, where Fig. 2.1(a) shows a constant wavelength for different amplitudes and Fig. 2.1(b) shows a fixed amplitude for different wavelengths. From these figures one can see clearly that the pulse width of a cnoidal wave is dependent on both amplitude and wavelength, which is in contrast to the single parameter dependence of $A \cdot \text{sech}^2(\theta)$ in Sec. 2.1. In Fig. 2.1(b) we show the infinite wave length case, i.e., the solitary wave. Fig. 2.1(c) shows the solitary wave for different amplitudes, illustrating the change in pulse width with amplitude.

2.2.1 Asymptotics of the Cnoidal Wave

The cnoidal wave solution converges to the $A \cdot \text{sech}^2(\theta)$ solution, however, if we extend the wavelength, λ, to ∞. This occurs when $K(m) \to \infty$ or when $m \to 1$. For $m = 1$, z_1 must be equal to z_2 (see 2.21), or for the zero offset case $z_1 = z_2 = 0$. The result is a solitary wave of the form:

$$u(x,t) = z_3 \cdot \text{sech}^2 \left[\sqrt{\frac{z_3}{2}} (x - 2z_3 t) \right] . \tag{2.30}$$

By setting $z_3 = A$ we obtain the same solitary wave solution as in (2.2),

$$u(x,t) = A \cdot \text{sech}^2 \left[\sqrt{\frac{A}{2}} (x - 2At) \right] . \tag{2.31}$$

The cnoidal wave can also converge to a cosinusoidal wave as $m \to 0$. This occurs when the amplitude, z_3, becomes very small (from (2.18), $z_1 < z_2$; $z_2 = 0$, therefore $z_1 < 0$ and $(z_3 - z_1) > 0$),

$$\begin{aligned} u(x,t) &= z_3 \cdot \cos^2 \left[\sqrt{\frac{-z_1}{2}} (x - 2z_1 t) \right] \\ z_3 &\ll 1. \end{aligned} \tag{2.32}$$

This makes physical sense because, for a small amplitude, the term $6uu_x$ of the KdV is much smaller than the other terms due to its square dependence, allowing one to approximate the KdV as a simple linear dispersive system, with a cosinusoidal squared solution:

$$u_t + u_{xxx} = 0. \tag{2.33}$$

2.2.2 The Cnoidal Wave vs. the Periodic Sech^{2}Wave

For long wavelengths and large m the cnoidal wave appears similar to a periodic train of solitary waves. In fact, we can approximate the cnoidal wave by a periodic train of sech^2 pulses. Figure 2.2 shows the overlay of a cnoidal wave with a periodic train of sech^2 solitary waves:

$$cn^2[x - 3; .98]; \ m = 0.98 \tag{2.34}$$

$$\sum_{n=0}^{\infty} \text{sech}^2[x - (3 + 6.71 \cdot n)]. \tag{2.35}$$

They match very well except for the slight error in between the pulses. By equating the pulse widths and velocities of (2.2) and (2.23) we find that for $m > 0.98$ the pulse width error is $< 1\%$ and the velocity error is $< 2\%$. The work presented in this book will be almost exclusively in the regime of $m > 0.98$ and, therefore, the cnoidal wave solutions will be approximated as a train of sech^2 pulses for simplicity.

2.3 Transient Dynamics of the KdV

In the previous two sections we looked at cnoidal and solitary wave solutions of the KdV equation. We have not yet called these solutions *solitons*, however, as we have not shown yet that they possess the properties of solitons. In this section we will show the qualitative properties of these waves that led Zabusky and Kruskal to coin the name *soliton* [2].

2.3.1 Solitary Wave $\neq$ Soliton

The distinction between solitary waves and solitons is not a trivial one, since there are several equations that produce hyperbolic secant solitary wave solutions that are *not* solitons. Two examples are shown below:

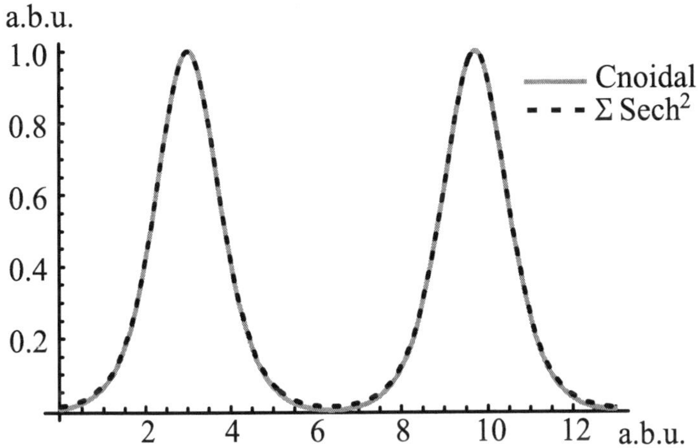

FIGURE 2.2
Cnoidal wave versus periodic sech[2] wave.

$$\frac{\partial}{\partial t}A = \alpha_1 A + \alpha_2 \frac{\partial^2}{\partial t^2}A + \alpha_3 |A|^2 A \tag{2.36}$$
$$b_t = -b_x + 6bb_x + b_{txx}. \tag{2.37}$$

The first is the master equation for optical modelocking (*non*-soliton) using a fast saturable absorber [45]. The solitary wave solution propagates through a *linear* medium and is shaped by a non-linear saturable absorber and a low-pass filter to create the sech shaped pulse[1]. The second is the regularized long wave equation [46], which also yields a sech shaped solitary wave solution. This solitary wave solution is not robust to collisions, a key property of solitons, and therefore the solitary wave solutions are not solitons.

These two examples illustrate the fact that the hyperbolic secant shape of a solitary wave does not make it a soliton. In Sec. 3.6 we will formalize a definition for a soliton based on this and the following two sections.

[1]Optical mode-locking can also create optical solitons when *dispersion* is used instead of a low-pass filter. The resulting solitons are based upon the nonlinear Schrödinger equation, which are different than the KdV soliton (See Ch. 1).

2.3.2 Soliton Collisions

One of the most remarkable properties of solitons is their ability to interact with one another nonlinearly and still maintain their identity after the interaction. These interactions occur naturally in KdV soliton systems since solitons have amplitude-dependent velocity, $c(A)$:

$$\begin{aligned} u(x,t) &= A \cdot \mathrm{sech}^2\left[\sqrt{\frac{A}{2}}(x-2At)\right] \\ c(A) &= 2A. \end{aligned} \tag{2.38}$$

In Fig. 1.1 we showed the qualitative results of an example collision. If a taller soliton is placed behind a shorter soliton it will catch up and collide with the smaller soliton. After the collision, however, the two solitons will reemerge fully intact, e.g., same amplitude and shape, and continue propagating at their original velocities. During the collision both solitons undergo a nonlinear change in shape which is distinctively different than the simple linear superposition of the two solitons.

Lax [29] showed that there are three cases of interest, defined by the relative speeds of the colliding solitons, c_1 and c_2:

$$\begin{array}{lrcl} \text{Case I:} & \frac{c_1}{c_2} & < & 2.62 \\ \text{Case II:} & 2.62 < & \frac{c_1}{c_2} & < 3 \\ \text{Case III:} & 3 & < & \frac{c_1}{c_2}. \end{array}$$

We will investigate each of these cases to illustrate the fascinating dynamics of soliton collisions.

The three cases are shown in Figs. 2.3-2.5. Note that in these figures the use of "Soliton 1" and "Soliton 2" and the trajectories are to help describe the qualitative picture for the reader. There is no way to discern between one soliton and the other during collision. Hence, one cannot draw a continuous trajectory for a soliton through the collision; nor can one say that the shorter soliton after collision was the taller one before collision, etc. One can only identify the taller and shorter soliton well before and well after the collision; there is no definitive means to connect, or identify, solitons before and after a collision.

- **Case I:** During the collision, the two solitons appear to interchange their roles without passing through one another [29]. This has been described as a collision of two tennis balls, where the faster ball collides with the slower ball and transfers its energy to the slower ball, causing the slower ball to accelerate and move away [4]. There are always two distinctive peaks during this collision.

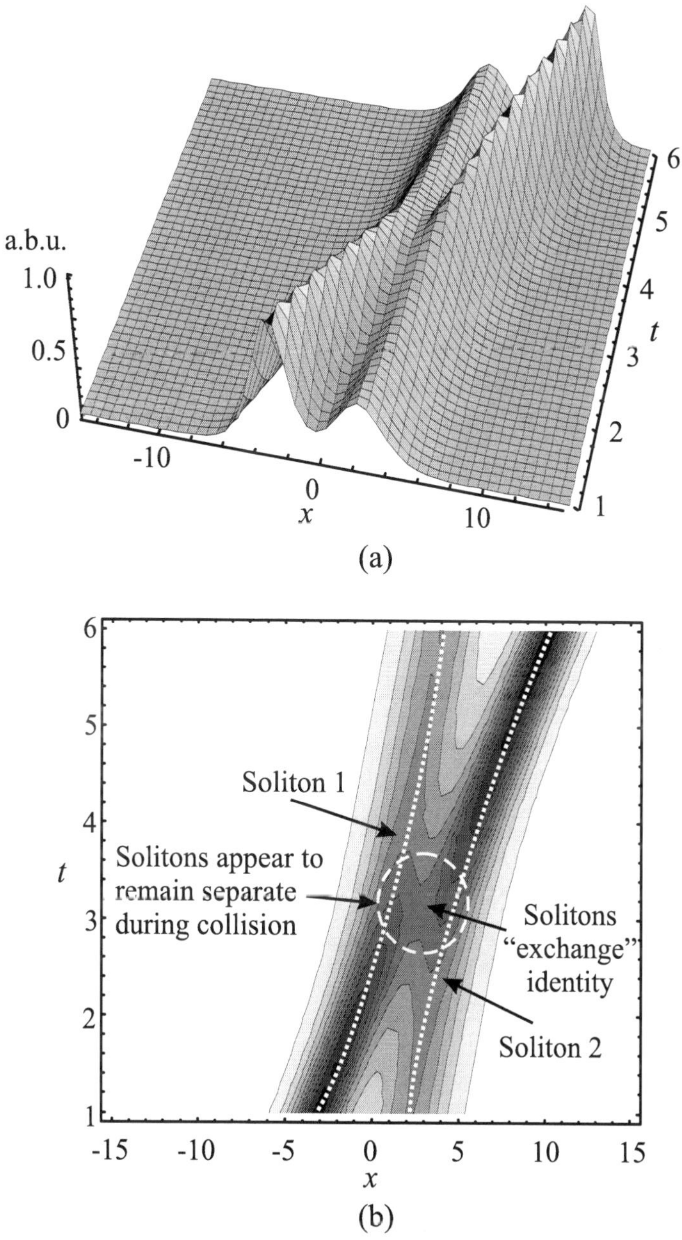

FIGURE 2.3
Soliton collision: Lax case I; velocity ratio: 2.25. (a) Soliton collision: the taller soliton collides with the shorter soliton and exchanges energy; the smaller soliton then grows and moves away. (b) Contour plot of collision; solitons appear to *not* pass through one another.

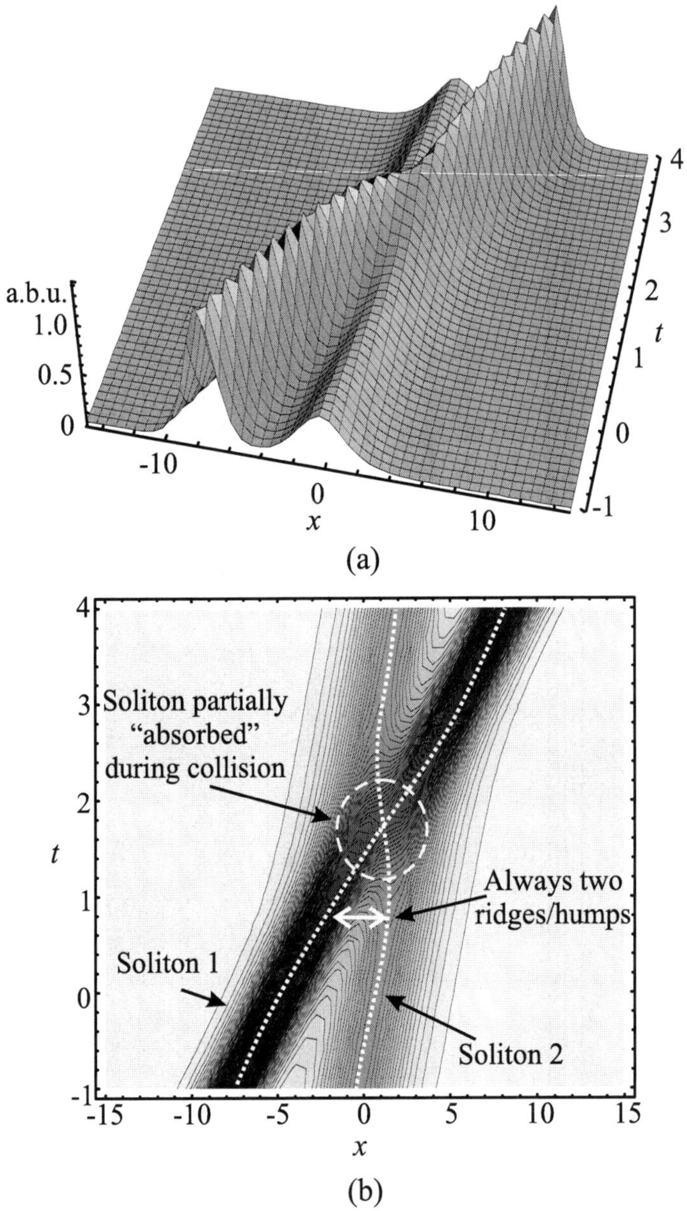

FIGURE 2.4
Soliton collision: Lax case II; velocity ratio: 2.82. (a) Soliton collision: the taller soliton collides with the shorter soliton and partially "absorbs" the smaller soliton; there is always an additional bump or peak inside the larger soliton as it absorbs the smaller soliton. (b) Contour plot of collision; two ridges are always present.

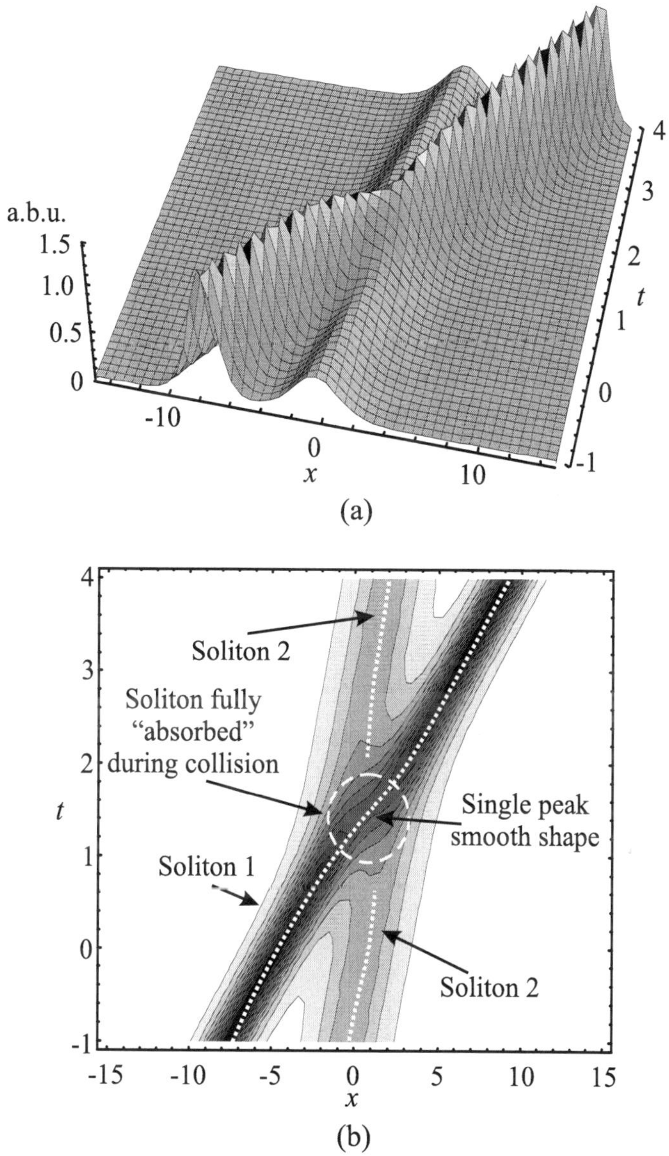

FIGURE 2.5
Soliton collision: Lax case III; velocity ratio: 3.06. (a) Soliton collision: the taller soliton collides with the shorter soliton and fully "absorbs" it; during collision there is a point where there is a single peak and a smooth surface. (b) Contour plot of collision; the smaller soliton "disappears" inside of the taller soliton during collision.

- **Case II:** During the collision, the taller soliton appears to absorb the smaller one and then re-emit it [29]. In this case there are always two distinctive peaks or bumps — during "absorption" there is always an extra peak in the large wave.

- **Case III:** During the collision, the taller soliton appears to absorb the smaller one and then re-emit it [29]. In this case there exists a point where there is only a single peak — the shorter soliton has been effectively "fully absorbed."

In all cases we observe two key properties of these collisions:

- **Solitons undergo nonlinear amplitude modulation during collision.**

- **Solitons incur a permanent phase modulation due to the collision.**

Figure 2.6 shows these properties explicitly for Case III above. In this figure, we plotted the soliton waveforms in gray and superimposed a dashed waveform which represents the linear superposition of two $\text{sech}^2(x)$ pulses traveling at the same velocities as the two solitons (the taller one is faster than the shorter one)[2]. Figure 2.6 (a) and (b) show the phase shift in soliton trajectory, ϕ_κ, due to collision. We see clearly in Fig. 2.6(b) that the larger soliton (gray) has been advanced and the shorter soliton (gray) has been retarded in comparison to the linear superposition of two $\text{sech}^2(x)$ pulses (dashed). In Fig. 2.6 (c) we see a close-up of the collision, where the nonlinear amplitude modulation of two solitons colliding (gray) can be easily contrasted against the linear superposition of one $\text{sech}^2(x)$ pulse overtaking another (dashed). Note that the dashed waveform is not a physical one — it is just a mathematical construct for comparison.

2.4 Summary

In this chapter we have presented the fundamental derivations of the KdV soliton and its steady-state solution. Figures 2.3–2.6 begin to show us the beauty of solitons and hint at a deeper underlying structure. The next chapter explores this structure and the underpinning of soliton dynamics.

[2]This was done by plotting two solitons independently and simply linearly adding amplitudes to create an artificial waveform that represents the linear superposition of two solitons.

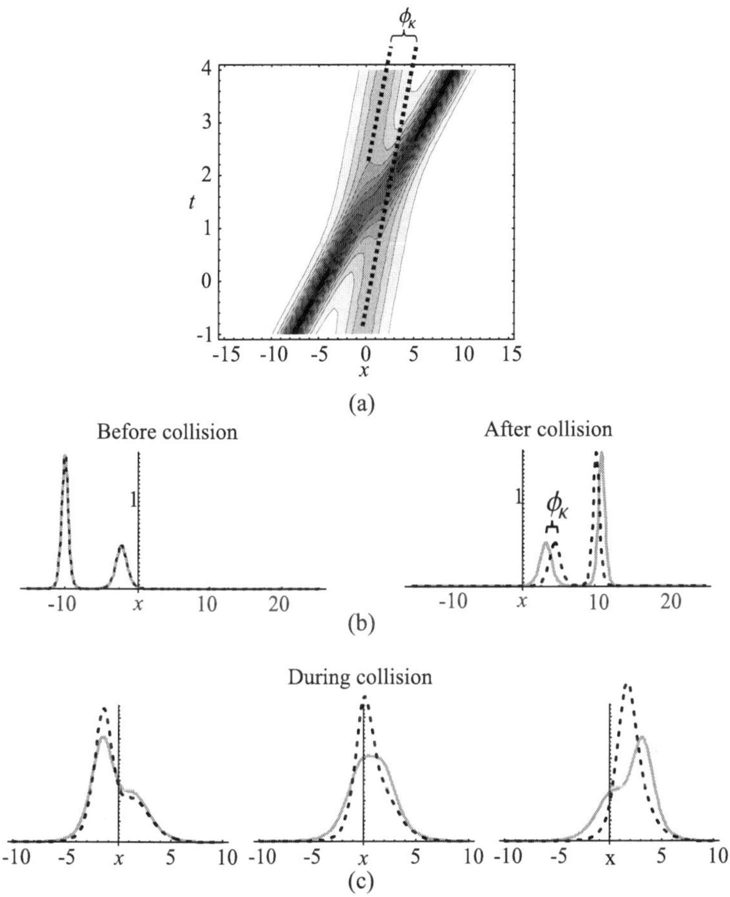

FIGURE 2.6
(a) Contour plot of soliton collision. Phase shift, ϕ_κ, is shown. (b) Before and after pictures of a soliton collision. Gray is KdV soliton collision; dashed is linear superposition of two sech^2 pulses traveling at the same velocity as the solitons (taller pulse faster than shorter pulse). (c) Close-up of soliton collision. Soliton collision (gray) differs significantly from superposition of two sech^2 pulses (dashed).

2.A Elliptic Integrals, Functions and Their Link to Differential Equations

We start with the motivation for using elliptic integrals and elliptic functions by examining a class of nonlinear, polynomial differential equations:

$$\left(\frac{dx}{dt}\right)^2 = C_3 \cdot x^3 + C_2 \cdot x^2 + C_1 \cdot x + C \tag{A-1}$$

These are nonlinear differential equations that have the solution, $u(x)$, of the form

$$u(x) = \int_0^x \frac{dy}{\sqrt{(y - z_1)(y - z_2)(y - z_3)}}. \tag{A-2}$$

This form is common in several geometric problems. Two examples are the arclength of a circle, Fig. A.1(a) and an ellipse, Fig. A.1(d).

2.A.1 Arclength of a Circle

To find the arclength of a circle we employ the arclength formula, which is a common formula found in many calculus texts for finding the arclength of an arbitrary curve. For a curve parameterized as $X = f(m)$ and $Y = f(m)$, where m is simply some parameter, the arclength, δ, is given as

$$\delta = \int_a^b \sqrt{\left(\frac{dX}{dm}\right)^2 + \left(\frac{dY}{dm}\right)^2}\, dm. \tag{A-3}$$

A circle of radius 1 can be defined as

$$x^2 + y^2 = 1,$$

or in a parameterized form

$$\begin{aligned} X(m) &= \sqrt{1 - m^2}, \\ Y(m) &= m. \end{aligned}$$

This yields the arclength

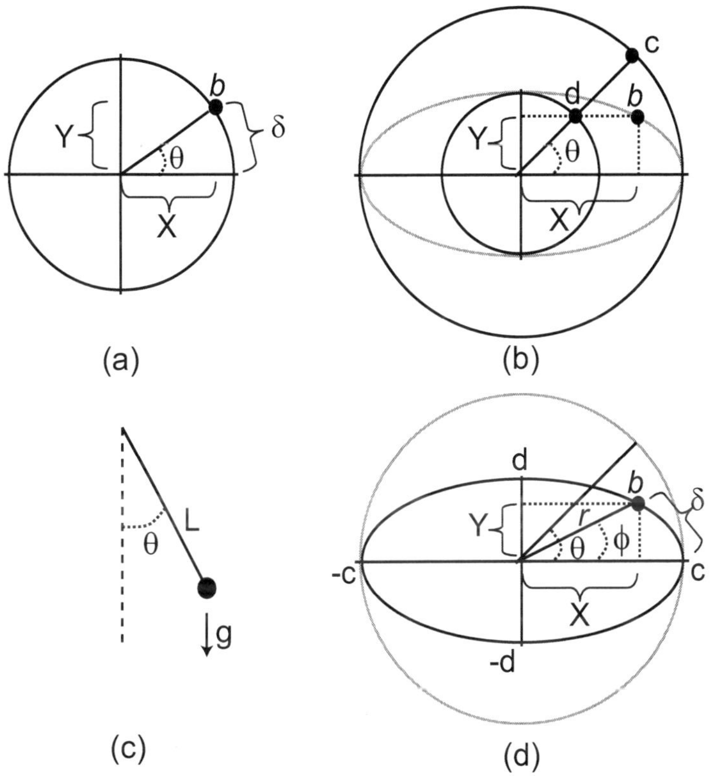

FIGURE A.1
(a) Arclength of a circle. (b) Parametrization of an ellipse with two concentric circles of radius c and d: $X = c \cdot \cos\theta$ and $Y = d \cdot \sin\theta$ (c) Nonlinear pendulum. (d) Arclength of an ellipse. Angle between line $\vec{r}$ and x axis is ϕ. Note $\phi \neq \theta$.

$$\delta = \int_a^b \sqrt{\left(\frac{-m}{1-m^2}\right)^2 + (m)^2}\, dm,$$
$$\delta = \int_a^b \frac{1}{\sqrt{1-m^2}}\, dm. \tag{A-4}$$

Given that for a circle of radius 1, the arc length is equal to the angle, θ, we can write this as

$$\arcsin(y) = \delta = \theta, \tag{A-5}$$
$$\arcsin(y) = \int_0^y \frac{1}{\sqrt{1-m^2}}\, dm, \tag{A-6}$$

which is a common definition of arcsin. We begin to see the connection between certain geometries and integrals of the form (A-2).

2.A.2 Arclength of an Ellipse

We now examine an ellipse, which we may write as

$$\frac{x^2}{c^2} + \frac{y^2}{d^2} = 1, \tag{A-7}$$

with eccentricity, e, defined as

$$\frac{\sqrt{c^2 - d^2}}{c}. \tag{A-8}$$

This can be written in parametric form as

$$X = c \cdot \sin(\theta) \quad , \quad \frac{dX}{dm} = c \cdot \cos(\theta) \tag{A-9}$$
$$Y = d \cdot \cos(\theta) \quad , \quad \frac{dY}{dm} = -d \cdot \sin(\theta). \tag{A-10}$$

where θ is defined in Fig. A.1(b) and is a parameter like m was for our circle. Note that, geometrically, θ is *not* the angle between $\vec{r}$ and the x axis, but rather the angle created by a line from the origin to two concentric circles of radii c and d.

Using this parametrization, the arclength, δ, is

$$\delta = \int_0^b \sqrt{c^2\cos^2(\theta) + b^2\sin^2(\theta)}\, d\theta \tag{A-11}$$

or, using the eccentricity, $e = k$,

$$\delta = c\int_0^b \sqrt{1 - k^2\cdot\sin^2(\theta)}\, d\theta, \tag{A-12}$$

which is the elliptic integral of the second kind.

2.A.3 Elliptic Functions

We can also examine this same problem in terms of the angle, ϕ, which is the angle between the line $\vec{r}$ and the x axis. This is similar in concept to the angle θ in the circular problem of Fig. A.1(a). This derivation was inspired, in part, by the course notes of Prof. William Schwalm at the University of North Dakota on elliptic functions.

We begin by rewriting the results above for an ellipse in terms of ϕ instead of θ. The angle between $\vec{r}$ and x-axis can be written in terms of θ as follows:

$$\begin{aligned}
&\phi = \arctan\frac{y}{x}\\
&\Rightarrow \tan\phi = \frac{y}{x}\\
&\Rightarrow \tan\phi = \frac{d}{c}\tan\theta\\
&\Rightarrow (1+\tan^2\phi)d\phi = \frac{d}{c}(1+\tan^2\theta)d\theta\\
&\Rightarrow \left(1+\frac{d^2}{c^2}\tan^2\theta\right)d\phi = \frac{d}{c}(1+\tan^2\theta)d\theta\\
&\Rightarrow d\phi = \frac{d/c}{\cos^2\theta + \frac{d^2}{c^2}\sin^2\theta}d\theta
\end{aligned}$$

To motivate the elliptic functions from the geometry of the ellipse we now integrate $\vec{r}$ directly along the ellipse using angle ϕ:

$$\begin{aligned}
u &= \int_0^{\phi} r d\phi \\
&= \int_0^{\phi} \sqrt{X^2 + Y^2} d\phi \\
&= \int_0^{\theta} \sqrt{c^2 \cos^2\theta + d^2 \sin^2\theta} \frac{d/c}{\cos^2\theta + \frac{d^2}{c^2}\sin^2\theta} d\theta \\
&= d \int_0^{\theta} \frac{d\theta}{\sqrt{\cos^2\theta + \frac{d^2}{c^2}\sin^2\theta}} \\
&= d \int_0^{\theta} \frac{d\theta}{\sqrt{1 - \left(1 - \frac{d^2}{c^2}\right)\sin^2\theta}}
\end{aligned}$$

with again $\theta = \arctan[(c/d)\tan\phi]$. For $d = 1$, and $k^2 \equiv 1 - d^2/c^2$, this becomes

$$u = \int_0^{\theta} \frac{d\theta}{\sqrt{1 - k^2 \sin^2\theta}} \tag{A-13}$$

which is just the elliptic integral of the first kind.

Finally, the Jacobi elliptic functions can be defined using

$$\begin{aligned}
\text{sn } u &= \sin\theta \\
\text{cn } u &= \cos\theta \\
\text{dn } u &= r/c
\end{aligned}$$

We note that for a circle, $c = d$ and $k = 0$, (A-13) reduces to $\text{sn}(u) = \sin(u) = \sin(\theta)$ and for $k = 1$ it reduces to $\text{dn}(u) = \tanh(u) = \tanh(\theta)$; thus demonstrating the connection of elliptic functions to standard and hyperbolic trigonometric functions.

2.A.4 Example PDEs

In terms of our original goal of solving nonlinear, polynomial equations we note that the solution to the KdV equation can be written in this form [See (2.12)–(2.20)]

$$\frac{1}{2}(u_\xi)^2 = -u^3 + \frac{c}{2}u^2 + C_1 u + C_2,$$

yielding the integral

$$\int_{u_o}^{u} \frac{du}{\sqrt{2}\sqrt{-(u-z_1)(u-z_2)(u-z_3)}} = \int_{\xi_o}^{\xi} d\xi,$$

and solution, in terms of the elliptic function cn

$$\begin{aligned} u(x,t) &= z_2 + (z_3 - z_2) cn^2 \left[\sqrt{\frac{z_3 - z_1}{2}} \xi; m \right], \\ m &= \frac{z_3 - z_2}{z_3 - z_1}. \end{aligned} \tag{A-14}$$

Other physical problems also find this form, such as the nonlinear pendulum in the large amplitude limit, Fig. A.1(c). The governing equation can be written as

$$\frac{d\theta^2}{dt^2} = -\frac{g}{L}\sin(\theta), \tag{A-15}$$

with solution

$$\begin{aligned} \theta(t) &= 2 \arcsin \left[k \cdot \mathrm{sn} \left(\frac{\mathrm{g}}{\mathrm{L}} \mathrm{t}, \mathrm{k} \right) \right], \\ k &= \sin(\theta_{max}/2). \end{aligned} \tag{A-16}$$

These examples show clearly connection of elliptic integrals and functions with nonlinear, polynomial differential equations.

3

The Heart of the Soliton: Inverse Scattering

CONTENTS

In the previous chapter we looked at the quantitative solutions for steady-state solitons as well as the qualitative transient behavior of solitons during collision. It is their unique properties *during* collision that give us a glimpse into the deep theoretical and mathematical underpinning of the soliton. In this chapter we provide an overview of the underlying structure of solitons that is embodied in a mathematical method known as inverse scattering. This technique not only will provide us with a quantitative description of transient soliton dynamics, it will guide us in understanding soliton collisions and the re-emergence of solitons after a nonlinear collision. Moreover, the inverse scattering method will allow us to determine the number and amplitude/speed/width of solitons that emerge from an arbitrary initial condition. This is extremely important for practical applications where a non-soliton initial condition is used, e.g., pulse sharpening.

3.1 Inverse Scattering Method

Soon after the discovery of the soliton, researchers began to examine the generation of solitons from an initial profile or disturbance. This initial value problem was only examined experimentally, however, since no analytical method was available to evaluate the nonlinear KdV equation. In the 1950s, the advent of the computer allowed researchers to begin to examine the dynamics of solitons more fully, including the elusive initial value problem. It is from these studies that the first glimpse was seen of the deep underlying properties of soliton systems. A framework for these underlying dynamics didn't come until 1967, however, when an amazing technique was developed by Gardner, Greene, Kruskal and Miura at Princeton [28] known as the inverse scattering method (ISM). The backbone of the technique was formalized a year later by Lax [29] when he showed that the operators chosen by Gardner and colleagues formed a Lax pair and that there existed an infinite number of nonlinear equations with these unique soliton properties (the KdV is the only one of physical and practical interest to us).

The unfortunate aspect of this amazing technique is that it involves some concepts and calculations that are neither trivial nor obvious. In fact, to many it appears some parts are simply pulled out of the air. In order to remove some of the mystery of the technique we will cover some of the very basic points of the underlying principles before we move on to look at the mechanics of finding a solution from an initial profile.

3.2 A Math Problem

We begin by examining a familiar math problem that has apparently nothing to do with our goal of solving the KdV equation, the linear eigenvalue problem. We assume a linear operator, L, that is composed of derivatives ($\frac{\partial^n}{\partial x^n}$) and multiplicative constants and has some form of dependence on a function, which we will call $u(x,t)$. The operator has no other time dependence except through $u(x,t)$. Right now, $u(x,t)$ is just some function, however we will see shortly that it relates directly to the soliton which we have also written as $u(x,t)$ elsewhere in this book. We write the eigenvalue problem as follows:

$$L_{[u(x,t)]}\psi = \lambda\psi, \tag{3.1}$$

where we have noted explicitly the dependence of L on $u(x,t)$ as $L_{[u(x,t)]}$. Note that in *this* eigenvalue equation x is the variable and t is a parameter, i.e., a number, such that for a given t we can solve this eigenvalue problem for λ. In other words, we do not solve for λ as a function of time, but rather for a specific time, t. Both λ and ψ may change with time, t, however, since the operator L is dependent on time via $u(x,t)$.

Let us assume that ψ evolves according to:

$$\psi_t = M\psi, \tag{3.2}$$

where M is some unknown operator. Now, let us look for the change in λ with time by taking the derivative of (3.1) with respect to time:

$$L_t\psi + L\psi_t \quad = \quad \lambda_t\psi + \lambda\psi_t. \tag{3.3}$$

Using (3.2) and (3.1) this can be rewritten as[1]:

$$\begin{aligned} (L_t - ML + LM)\psi &= \lambda_t\psi \\ (L_t - [M,L])\psi &= \lambda_t\psi \end{aligned} \tag{3.4}$$

From (3.4) we see that λ_t is zero if:

$$L_t \quad = \quad [M,L] \tag{3.5}$$

This is an amazing result, as it says that the eigenvalues do not change with time if (3.5) is satisfied. In addition, if M is a simple enough operator, we can find $\psi(t)$ for all t from $\psi(0)$ via (3.2). Let us now return to the eigenvalue problem at some time t ($\lambda(t) = \lambda(0) = \lambda$):

$$L_{[u(x,t)]}\psi(t) = \lambda\psi(t). \tag{3.6}$$

We know everything in (3.6) except $u(x,t)$ in L. Since L is a linear

---

[1] $[A,B] \equiv AB - BA$

operator and (3.6) a linear equation, it might be possible to determine $u(x, t)$ by solving the inverse of the eigenvalue problem. In other words, we could determine the operator L, whose only unknown is $u(x, t)$, from the eigenvalues, λ, and the eigenfunctions, $\psi(t)$. It is not clear at this point, however, that such an "inverse" solution is achievable. It should be noted that the "win" in this approach is that instead of solving the evolution of $u(x, t)$ in time from $t = 0$, we solve $\psi(t)$ from $\psi(0)$ using (3.2). This may seem pointless, unless M is a very simple operator and the evolution of $u(x, t)$ is very complicated. For example, perhaps M is a linear operator where as the evolution of $u(x, t)$ is defined by a nonlinear equation. We shall see shortly, that for certain systems, this is indeed the case.

3.3 KdV Solution via the Inverse Scattering Method

So what does this have to do with solving the KdV equation? And how do we find the operators L and M? The answer to the first question lies in the $u(x, t)$ dependence of L. The answer to the second came from Gardner and colleagues [28] in a brilliant intuitive selection of L as the Schrödinger operator:

$$L_{[u(x,t)]} = \frac{\partial^2}{\partial x^2} + u(x, t), \tag{3.7}$$

where $u(x, t)$ evolves according to the KdV equation:

$$u_t + 6uu_x + u_{xxx} = 0. \tag{3.8}$$

With this L now chosen, one can look for an M that satisfies (3.5). Both Gardner and colleagues [28] and Lax [29] showed that:

$$M = 4\frac{\partial^3}{\partial x^3} + 3u\frac{\partial}{\partial x} + 3u_x. \tag{3.9}$$

We can now show (3.5) is true as follows. We rewrite the LHS of (3.5) as:

$$
\begin{aligned}
L_t &= \frac{\partial}{\partial t}\left[\frac{\partial^2}{\partial x^2} - u(x,t)\right] \\
&= -u_t
\end{aligned} \tag{3.10}
$$

and the RHS as:

$$
[M, L] = 6uu_x + u_{xxx} \tag{3.11}
$$

which when combined gives (3.5):

$$
L_t - [M, L] = u_t + 6uu_x + u_{xxx} = 0 \tag{3.12}
$$

$$
L_t = [M, L]. \tag{3.13}
$$

This is an important result:

If $u(x,t)$ evolves according to the KdV equation, then (3.4) tells us that the eigenvalues of (3.1) are time independent!

In fact, we will see shortly, that the eigenvalues represent the individual solitons — each soliton can be represented by its own, time invariant eigenvalue. *This is the essence of the underlying dynamics of solitons.*

With L and M chosen we can now seek an inverse solution of (3.6) and through this a solution for $u(x,t)$ for all time. This method is shown graphically in Fig. 3.1 for L equal to the Schrödinger operator.

We pause a moment in order to make several important points:

- L and M are both **linear** operators, and thus (3.6) and (3.2) are likely solvable. $u(x,t)$ is treated as a parameter in these operators, not as a function: at time t, u is only a function of x.

- The "parameter," $u(x,t)$, evolves according to the KdV, which *is* a nonlinear differential equation.

- The inverse solution at time t is greatly helped by the fact that the eigenvalues do not change with time and ψ evolves according to the linear equation, (3.2). These properties are a result of the relationship: $L_t = [M, L]$, which is known as a Lax pair.

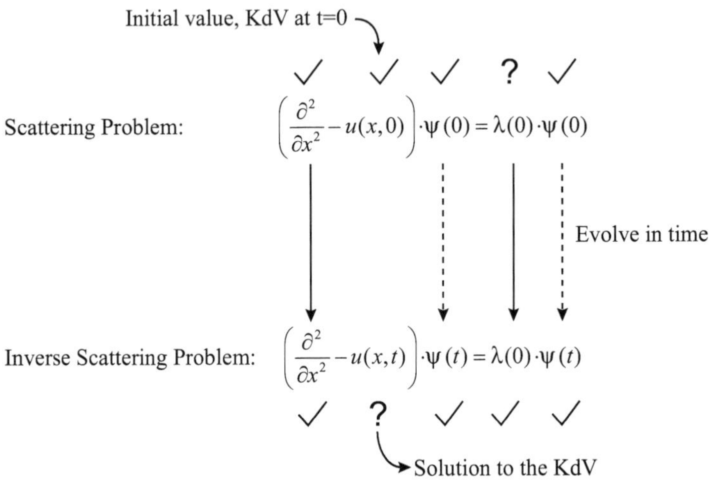

FIGURE 3.1
Graphical representation of the inverse scattering method. We start by solving the traditional scattering problem of the time-independent Schrödinger equation for the scattering data, or eigenvalues, $\lambda(0)$, using the initial potential profile $u(x, 0)$. We then propagate the eigenfunctions forward in time as shown by the dashed lines. The solid line represents time invariant quantities, e.g., λ. We now have an equation at time t. We know the eigenvalues and eigenfunctions, but not the potential shape, $u(x, t)$. We solve the inverse of the scattering problem to find the potential shape, which is the solution to the KdV equation at time t for an initial profile of $u(x, 0)$.

We now return to the inverse solution of (3.6). It is not a trivial by any means; however, Gel'fand and Levitan showed in 1951 that a solution is possible [47]. The inverse solution is found by first determining the eigenvalues and eigenfunctions at $t = 0$ and then propagating them forward in time (note $\lambda(t) = \lambda(0)$). The eigenfunctions, $\psi(t)$, at time t are given by,

$$\psi_t = M\psi. \tag{3.14}$$

For the M in (3.9) we note that as $x \to \infty$, M reduces to the simple form,

$$M = 4\frac{\partial^3}{\partial x^3}, \tag{3.15}$$

since for the soliton, u and u_x go to zero as $x \to \infty$. The inverse solution is found by solving for $K(x, z; t)$ in the Gel'fand-Levitan equation (see [3, 5, 19] for details):

$$K(x, z) + B(x + z) + \int_0^\infty K(x, y)B(y + z)\,dy = 0. \tag{3.16}$$

where the function, $B(x + z; t)$, is derived from the eigenvalues and eigenfunctions of (3.1), at time t (z is a dummy variable) [3, 19].

With $K(x, z; t)$ determined, the function $u(x, t)$ at time t can be computed as follows:

$$u(x, t) \quad = \quad 2\frac{\partial}{\partial x}K(x, x; t). \tag{3.17}$$

From (3.16), we can see that this process is by no means simple, and in fact it is only for a very simple set of discrete eigenvalues that it can be solved. Luckily for us, we are primarily interested in this case. The next section will describe the mechanics of finding the solution to an initial value problem using the *results* of the inverse scattering method for some common initial profiles. The reader should note that we will *not* go through the process outlined in this section, particularly the solution to (3.16), since an intuitive understanding and practical application do not require it. The full derivations can be found in [3, 5, 48].

3.4 Solution of the KdV Initial Value Problem

We begin by recognizing that (3.1) with (3.7) is the familiar scattering problem studied in quantum mechanics, where $u(x,t)$ corresponds to a potential. It is helpful to use this analogy since the operator, (3.7) and eigenvalue equation, (3.1), are encountered so often in the context of quantum mechanics that it is helpful to discuss the solutions in terms of the quantum mechanical problem. Inverse scattering, however, does not have any relation, whatsoever, to the quantum mechanical problem that the terminology is borrowed from. The only connection is similar mathematics.

3.4.1 The Eigenvalue Problem Using the Schrödinger Operator

We briefly solve (3.1) using (3.7) to show the how the eigenvalues are found and also to introduce the terminology commonly used, i.e., borrowed from quantum mechanics.

We begin by assuming an initial profile of $u(x, t = 0)$. For mathematical simplicity and without loss of generality, we will select a delta function, $\delta(x)$, Fig. 3.2. In terms of the quantum mechanical problem, $|\psi|^2$ represents the probability density function of an electron, i.e. the probability of finding an electron in space x. Since the overall probability must equal 1, the integral of $|\psi|^2$ over x is equal to 1. We now will solve the Schrödinger eigenvalue problem with this $u(x, t = 0)$ as our *potential*. The term potential refers to the context that the eigenvalues are related to the energy of the electrons in the system and $u(x)$ is the potential energy versus x of the physical system under consideration. The equation to solve is then,

$$L_{[u(x,t)]}\psi = \lambda\psi \tag{3.18}$$

$$-\frac{\partial^2\psi}{\partial x^2} - (\delta(x))\psi = \lambda\psi \tag{3.19}$$

We have modified L with two negative signs in order to coincide with the quantum mechanical analogy. The effect of this is that negative potentials will have discrete eigenvalues, or *bound* states. In electronics we think of positive initial conditions, i.e. $u(x, t = 0)$ is a positive waveform, whereas in quantum mechanics one often discusses potential wells, i.e., negative waveforms. The added negative signs create a potential "well,"

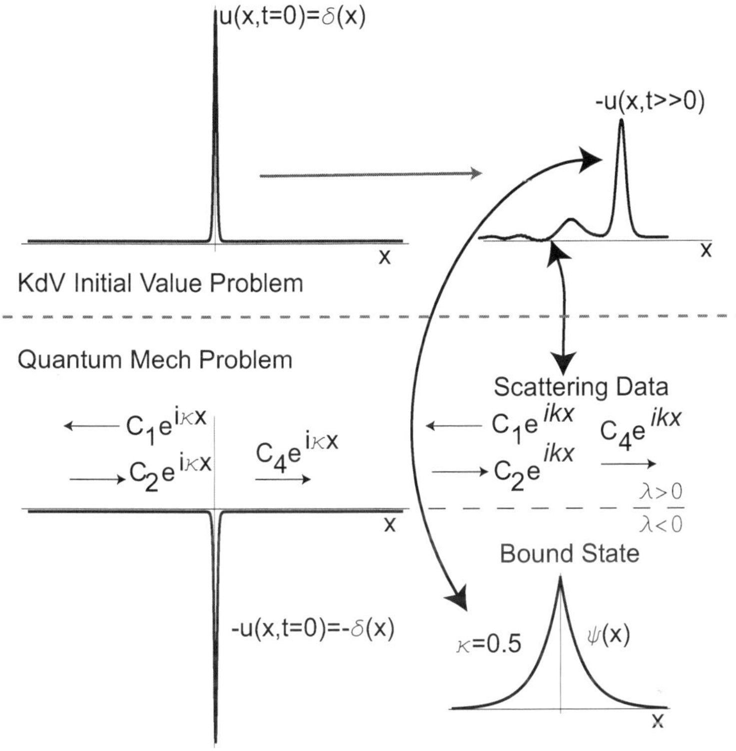

FIGURE 3.2
Top: delta function as initial condition at $t = 0$ forms into a single soliton and a dispersive tail. The soliton is determined by the eigenvalue of the bound state. Bottom: potential well for solving Schrödinger eigenvalue equation. Eigenfunction for bound state for $\lambda < 0$ and scattering data for $\lambda > 0$. The eigenvalue determines the soliton that emerges. The scattering data represent the dispersive tail, or non-soliton component of the solution.

Fig. 3.2, bottom, from our initial $u(x, t = 0)$, Fig. 3.2, top, and ensure discrete eigenvalues for that well.

We now solve (3.19). For regions $x < 0$ and $x > 0$, (3.19) simplifies to

$$-\frac{\partial^2 \psi}{\partial x^2} = \lambda \psi. \tag{3.20}$$

We now define

$$\kappa \equiv \sqrt{-\lambda}. \tag{3.21}$$

For negative λ, κ is real and results in "bound" states, or states that are localized in space. For positive λ, κ is imaginary and results in states that are not bound in space, i.e. waves.

We begin by considering negative λ. The solution to (3.20) is of the form

$$\psi(x) = C_1 e^{-\kappa x} + C_2 e^{\kappa x} \quad x < 0, \tag{3.22}$$
$$\psi(x) = C_3 e^{-\kappa x} + C_4 e^{\kappa x} \quad x > 0. \tag{3.23}$$

Using the property that the solution must not go to infinity at $|x| = \infty$, we can write the solution for (3.20), $x \neq 0$, as

$$\psi(x) = C_2 e^{\kappa x} \quad x < 0, \tag{3.24}$$
$$\psi(x) = C_3 e^{-\kappa x} \quad x > 0. \tag{3.25}$$

To find the complete solution of (3.19) we impose the following boundary conditions:

$$\begin{array}{ll} \psi & \text{is continuous} \\ \frac{\partial \psi}{\partial x} & \text{is continuous, except at discontinuities.} \end{array} \tag{3.26}$$

The first condition results in

$$C_2 = C_3 = C \tag{3.27}$$

The second requires us to examine the solution near the discontinuity (the delta function). We solve this by integrating (3.19) from $-\epsilon$ to ϵ and looking at the solution as $\epsilon \to 0$.

$$-\int_{-\epsilon}^{\epsilon} \frac{\partial^2\psi}{\partial x^2}\,dx - \int_{-\epsilon}^{\epsilon} \delta(x)\psi(x)\,dx = \kappa^2 \int_{-\epsilon}^{\epsilon} \psi^2(x)\,dx \tag{3.28}$$

$$-\frac{\partial\psi}{\partial x}\bigg|_{-\epsilon}^{\epsilon} - \psi(0) = \kappa^2 \int_{-\epsilon}^{\epsilon} \psi^2(x)\,dx. \tag{3.29}$$

The term on the RHS is zero since ψ is finite and when we take the limit at $\epsilon \to 0$, the area will be zero. This leaves us with

$$-\frac{\partial\psi}{\partial x}\bigg|_{-\epsilon}^{\epsilon} = \psi(0) \tag{3.30}$$

using (3.24) and (3.25) we obtain

$$2C\kappa = C \tag{3.31}$$

$$\kappa = \frac{1}{2}. \tag{3.32}$$

This result shows us that there is a **single** eigenvalue for a delta function potential. This is important, since each eigenvalue corresponds to a soliton, Fig. 3.2, right. The result is that a delta function initial waveform will always generate a single soliton.

We can find C by using the normalization condition: the integral of $|\psi|^2$ over x is equal to 1,

$$\int_{-\infty}^{\infty} |\psi(x)|^2\,dx = 2|C|^2 \int_{0}^{\infty} e^{-2\kappa x}\,dx = 1, \tag{3.33}$$

$$C = \frac{1}{\sqrt{2}} = \sqrt{\kappa}. \tag{3.34}$$

We now return to positive λ and define[2] $k = \sqrt{\lambda}$. The solution will be of the form

$$\psi(x) = C_1 e^{-ikx} + C_2 e^{ikx} \quad x < 0, \tag{3.35}$$

$$\psi(x) = C_3 e^{-ikx} + C_4 e^{ikx} \quad x > 0. \tag{3.36}$$

[2]This means that $\kappa = k \cdot i$

We cannot eliminate any of the solutions since all are bounded at $x = \infty$. Boundary conditions require that at $x = 0$

$$C_1 + C2 = C3 + C4 \tag{3.37}$$

and

$$-\frac{\partial \psi}{\partial x}\bigg|_{-\epsilon}^{\epsilon} = \psi(0), \tag{3.38}$$

with

$$\frac{\partial \psi}{\partial x} = ik(-C_1 e^{-ikx} + C_2 e^{ikx}) \;\; x < 0, \tag{3.39}$$

$$\frac{\partial \psi}{\partial x} = ik(-C_3 e^{-ikx} + C_4 e^{ikx}) \;\; x > 0, \tag{3.40}$$

resulting in

$$ik(-C_3 + C_4 - C_2 + C_1) = C_1 + C_2. \tag{3.41}$$

Unlike for $\lambda < 0$, we do not have enough information to solve for k (5 unknowns and two equations). If one assumes that the two solutions, (3.35) and (3.36), represent an incoming wave, $C_2 e^{ik}$, a wave reflected, or scattered, by the delta function potential, $C_1 e^{-kx}$, and a wave that passes by, or is transmitted through, the delta function potential, $C_4 e^{ikx}$, a solution in terms of the input can be derived (C_3 is zero since it cannot be generated from an C_2 input). To do this we use (3.37) and (3.41) to derive the relationship between the reflected and transmitted waves and the incoming wave,

$$C_1 = \frac{1}{2ki - 1} C_2, \tag{3.42}$$

$$C_4 = \frac{2k}{2k + i} C_2. \tag{3.43}$$

We can define the amount of the wave reflected and transmitted by looking at the ratio of the squares of the waves [3]

[3]The use of the square of the wave is because the square of the wave represents the probability density of an electron wavefunction and as such is a physical quantity we can constrain [49].

$$R = \frac{|C_1|^2}{|C_2|^2} = \frac{1}{4k^2+1}, \tag{3.44}$$

$$T = \frac{|C_4|^2}{|C_2|^2} = \frac{4k^2}{4k^2+1}. \tag{3.45}$$

These are important quantities since a waveform with only solitons will have $R = 0$ and $T = 1$, which we refer to as reflectionless potentials (See Subsec. 3.5.1).

R and T, together with the input, $C_2 e^{ik}$, represent the scattering data. It is important to notc that there are an **infinite** solutions when λ is positive. The *bounded states* (discrete eigenvalues) and the *scattering data* (continuous, or infinite, eigenvalues) (See Fig. 3.2) are the components used to derive $B(x+z;t)$ mentioned in the previous section[4].

In summary, we have derived the eigenvalues and eigenfunctions of the delta function potential.

3.5 Asymptotic Solution to the Inverse Scattering Method

In the previous section we showed the calculation of the discrete eigenvalues representing the bounded states. Using these discrete eigenvalues, **we can write the asymptotic solution of the solutions simply in terms of the square root of the discrete eigenvalues** $\sqrt{-\lambda} = \kappa$

$$u(t) = 2\kappa^2 \cdot \text{sech}^2[\kappa(x - 4\kappa^2 t + \phi)]! \tag{3.46}$$

This solution is valid for $t >> 0$, such that the transient soliton formation has already occurred (See Fig. 3.2). This solution omits the oscillatory tail of the non-soliton component and also does not tell us the phase, ϕ, of the soliton, i.e. where it is in time. The tail is of little interest to many applications which either ignore or attenuate this portion. The indeterminate phase is usually not a concern as we are concerned with the shape/characteristics of the emerging soliton only, not its specific location in space. If a complete solution is required, one must resort to *numerical* calculation since the inverse scattering method only provides

[4]The derivation is not shown here as it involves a more detailed treatment and is not useful in the final, practical application of the inverse scattering method in electrical solitons.

analytical solutions for systems with only discrete eigenvalues. If the transmission or reflection coefficients, T and R, are not 1 and 0, no analytical solution can be found. If multiple discrete eigenvalues are present and $T = 1$ and $R = 0$, a fully analytic solution may be found, including the phase information. This is accomplished by performing the complete inverse scattering calculation using the (3.16).

3.5.1 Reflectionless Potentials

In the Sec. 3.3 we noted that a closed-form solution of (3.16) is not possible for scattering data which contain a continuous spectrum [3]. We therefore focus only on potentials $[-u(x,0)]$ which have only bound states and no continuous scattering data, i.e. $T = 1$ and $R = 0$. Such potentials are transparent to the continuous spectrum and as such do not cause any reflections, hence the absence of any continuous scattering data. These potentials are called "reflectionless" potentials. The bound states inside the reflectionless potential can be expressed as a finite series of n discrete eigenvalues: $-\lambda_n = \kappa_n^2$. By following the process described in Sec. 3.3 and solving (3.16) with a $B(x+z;t)$ containing only discrete eigenvalues we can find a closed form solution for the initial value problem. This solution will tell us how the initial waveform evolves in time from the initial potential, $u(x,0)$, to a set of discrete solitons, $u(x,t')$, at time t'. The nth soliton that emerges is given by the nth eigenvalue of the bounded states of $u(x,0)$:

$$u_n(x,t) = 2\kappa_n^2 \cdot \operatorname{sech}^2(\kappa_n x - 4\kappa_n^3 t), (\kappa > 0) \tag{3.47}$$

$$\kappa_n = \sqrt{-\lambda_n} \tag{3.48}$$

By definition any waveform consisting of only solitons will be reflectionless, even if those solitons are colliding with one another. As such there is an infinite number of waveforms that are reflectionless. In practice however, we are concerned with initial conditions of an electrical system which are generally constrained to a simple initial waveform. There are two important classes of reflectionless potentials that we will consider. The first is the soliton shape itself, for which there emerges only one soliton with the same κ as the initial profile (κ can be any real, positive number, not necessarily an integer):

$$\begin{aligned} \text{Reflectionless Potential I:} \quad u(x,0) &= 2\kappa^2 \cdot \operatorname{sech}^2(\kappa x) \\ \text{Emerging Soliton:} \quad &= 2\kappa^2 \cdot \operatorname{sech}^2(\kappa x - 4\kappa^3 t). \end{aligned} \tag{3.49}$$

The second reflectionless potential is a specially shaped sech^2 potential, for which N solitons emerge with $\kappa_n = n$ and $n = 1, 2, \ldots,$ N:

$$\text{Reflectionless Potential II:} \quad u(x,0) = \left[N^2 + N + \tfrac{1}{4}\left(\tfrac{l^2-1}{l^2}\right)\right] \cdot \mathrm{sech}^2\left[\tfrac{x}{l}\right] \tag{3.50}$$

or with $l = 1$,

$$\begin{aligned} \text{Reflectionless Potential II:} \quad u(x,)) &= N(N+1) \cdot \mathrm{sech}^2(x) \\ \text{Emerging } n^{th} \text{ Soliton:} \quad &= 2\kappa_n^2 \cdot \mathrm{sech}^2(\kappa_n x - 4\kappa_n^3 t). \end{aligned} \tag{3.51}$$

As an example of the power of the inverse scattering method, we can solve the KdV equation using the ISM for an initial potential of the form $u(x,0) = 2(2+1) \cdot \mathrm{sech}^2(x) = 6 \cdot \mathrm{sech}^2(x)$, for which two solitons ($N = 2$) will emerge. The solution is given by Drazin [3] as:

$$u(x,t) = 12\frac{3 + 4\cosh(2x - 8t) + \cosh(4x - 64t)}{[3\cosh(x - 28t) + \cosh(3x - 36t))]^2} \tag{3.52}$$

as $t \to \pm\infty$ the two separate solitons are given by:

$$\begin{aligned} u(x,t) &= 2 \cdot \mathrm{sech}^2\left[x - 4t \pm \frac{1}{2}\log 3\right] \\ &\quad +8 \cdot \mathrm{sech}^2\left[2x - 32t \mp \frac{1}{2}\log 3\right] \qquad (3.53) \\ &= \Sigma_{\kappa=1}^2 2\kappa^2 \cdot \mathrm{sech}^2\left[\kappa x - 4\kappa^3 t + \phi_\kappa\right]. \qquad (3.54) \end{aligned}$$

Hence, we have an exact analytical solution for the evolution of an initial $u(x,0) = 6 \cdot \mathrm{sech}^2(x)$ profile as it develops into two distinct solitons. There is only one additional term which we have not mentioned, ϕ_κ. This is the phase shift that we spoke about in Sec. 2.3.2. In addition we find that (3.52) gives us the solution for t positive and t negative, such that our initial profile $6 \cdot \mathrm{sech}^2(x)$ at $t = 0$ is simply the profile that results from the two discrete solitons in (3.53) colliding at $t = 0$. Fig. 3.3 plots (3.52) for time before and after $t = 0$.

To illustrate the time invariance of eigenvalues for general soliton waveforms, Fig. 3.4 shows the time evolution of two solitons as they collide with one another. Plotted with each waveform are the eigenvalues numerically calculated directly from the soliton waveform using the Schrödinger operator along with the calculated eigenfunctions corresponding to each soliton solution. From Fig. 3.4 it is clear that:

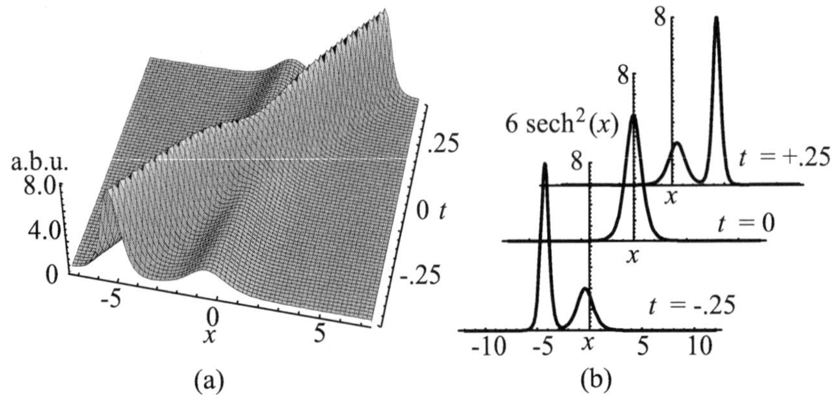

FIGURE 3.3
Drazin's solution for the evolution of a $u(x,0) = 6 \cdot \text{sech}^2(x)$ initial profile. Note that the solution given is for all time, such that we can also view the solution as the collision of two solitons that, at $t = 0$, form an exact $u(x,0) = 6 \cdot \text{sech}^2(x)$ shape.

- The eigenvalues do not change with time
- A waveform consisting of only solitons, regardless of position or collision, is reflectionless
- The eigenfunctions *do* evolve with time, as formulated in (3.14)

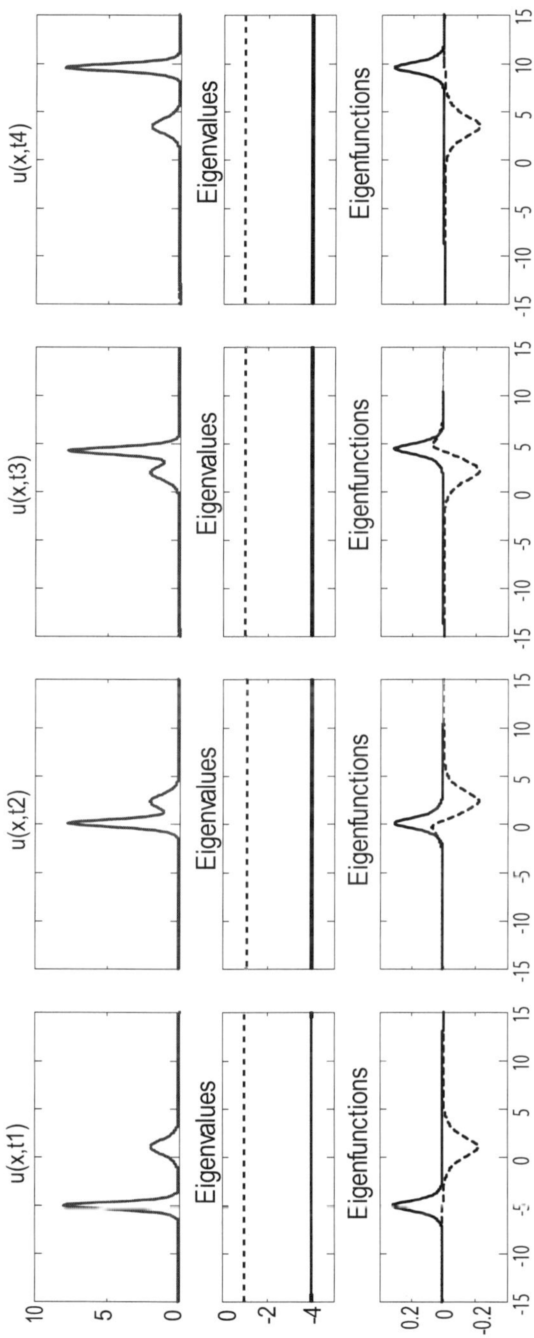

FIGURE 3.4
$u(x,t)$ of two solitons colliding at four different times. Shown with each $u(x,t)$ are the corresponding eigenvalues and eigenfunctions. Note that the eigenvalues are constant for all values of t and that there is no scattering data, i.e. the waveforms $u(x,t)$ are all reflectionless potentials. The eigenfunctions do evolve with time.

3.5.2 Non-Reflectionless Potentials

The above discussion of reflectionless potentials led us to an exact solution for the evolution of an initial profile. There are many cases, however, when we have a potential that is not reflectionless, and thus there are some continuous scattering data. As mentioned above, for such cases we cannot solve (3.16) for an exact analytical solution. We can, however, determine the asymptotic solution that emerges at some large time, t, later. The reason for this is that the continuous scattering data represent the non-soliton portion of the initial profile, i.e. the portion that does not form into a soliton. Numerical simulations show that this "excess" energy is shed in the form of a dispersive tail which propagates backwards in space, opposite from the propagation of the soliton(s). Thus, after a long enough time, the solitons have separated from their non-soliton component and are given simply by the bounded state eigenvalues as before. Figure 3.5 shows the numerical simulation of the evolution of a $4 \cdot \text{sech}^2(x)$ initial profile (non-reflectionless) into two solitons and a dispersive tail at three different times: $t = 0$, $t = .3$, and $t = .5$.

The asymptotic soliton solution (t large) of this non-reflectionless potential can be found by finding the bounded state eigenvalues for the initial profile, in this case $4 \cdot \text{sech}^2(x)$, which are given by (A is the amplitude, in our case $A = 4$)

$$\kappa_n = \sqrt{A + \frac{1}{4}} - (n - \frac{1}{2}) \tag{3.55}$$

$$\kappa_1 = 1.56 \tag{3.56}$$

$$\kappa_2 = 0.56. \tag{3.57}$$

Note that κ_n must be positive, so there are only two eigenvalues or two solitons that emerge.

Luckily the scattering problem defined by (3.1) with (3.7) is well known, and the solutions for many potential wells, or initial profiles, have already been found. For the general $A \cdot \text{sech}^2(x/l)$ potential well the nth eigenvalue ($\kappa^2 = \lambda$) is given as [48]:

$$\kappa_n = \frac{1}{2l}[(1 + 4Al^2)^{\frac{1}{2}} - (2n - 1)] \tag{3.58}$$

with,

$$\kappa - n > 0; \;\; n = 1, 2, 3, \ldots \tag{3.59}$$

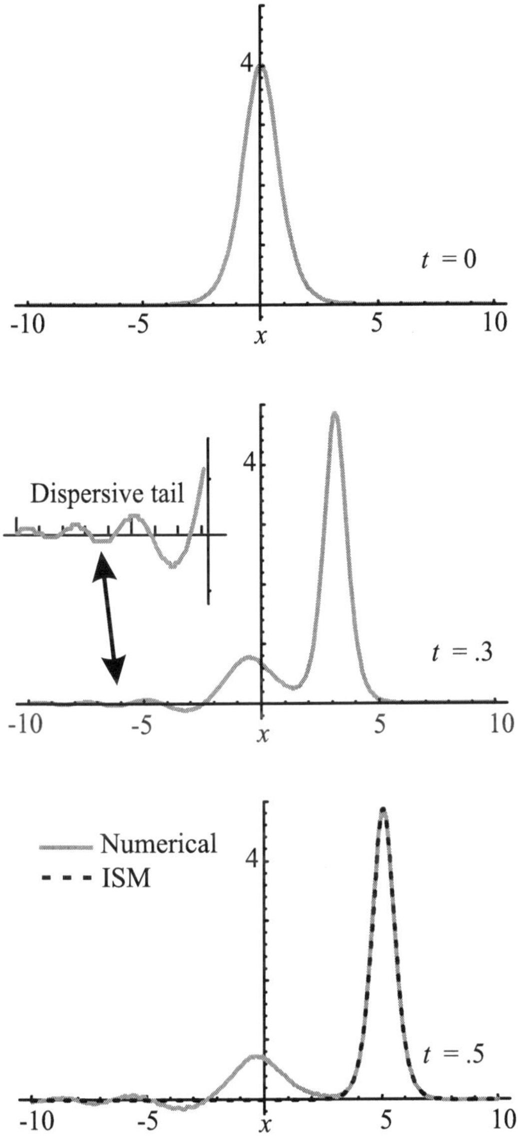

FIGURE 3.5
Non-reflectionless initial profile breaks up into soliton components and a dispersive tail. The solitons correspond to the bounded state eigenvalues of the scattering problem. The dispersive tail corresponds to the continuous spectrum of the scattering problem. At large time, t, the solution is simply the two discrete solitons represented by the bounded state eigenvalues of the initial profile. The dashed line is calculated from κ_1 and is aligned in time to overlay the numerical simulation (using the asymptotic method we have described, we do not know the phase of the emerging soliton).

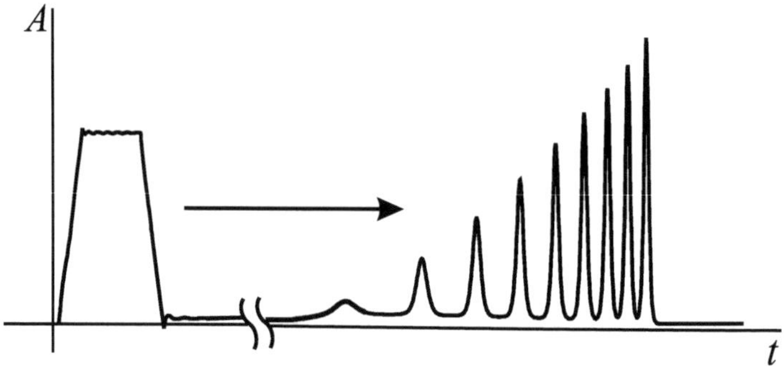

FIGURE 3.6
SPICE simulation: quasi-square shaped initial profile breaks up into multiple solitons, each one corresponding to a bounded state eigenvalue.

We see then that we can determine both the number, n, of solitons that emerge from a $A \cdot \mathrm{sech}^2(x/l)$ potential as well as the amplitude, width and velocity, via κ_n. Figure 3.5, bottom, shows the comparison between the soliton generated from an initial $4 \cdot \mathrm{sech}^2 x$ potential (numerical simulations, gray line) and one calculated via (3.58) (dashed line). Note that we do not have the phase information from the asymptotic solution of the ISM and, therefore, we do not know exactly where the soliton will be. In Fig. 3.5, bottom, we placed the soliton calculated from inverse scattering overtop the numerical one for comparison, i.e. we chose a phase that put the two pulses on top of one another. The ISM only tells us the amplitude and from that the width and velocity, the phase information is not found in the asymptotic solution.

The asymptotic solution for many other potential wells can also be determined, such as for a delta function (Sec. 3.4.1) and a finite square well. The eigenvalues for these cases have been well studied in the literature, which makes the calculations for the emerging solitons very simple. As an example, the breakup of a square pulse input into multiple solitons is shown in Fig. 3.6. Each of the emerging solitons represents a bounded-state of the initial well. We must emphasize, however, that for non-reflectionless potentials we can only calculate the asymptotic solution when the solitons have already formed.

3.6 Soliton Definition

In this chapter we have presented many of the characteristics of the soliton. We have not, however, defined exactly what a soliton is. Drazin proposes the following [3]:

"A soliton is that component of the solution of a nonlinear evolution equation which depends only upon one constant discrete eigenvalue (of the underlying scattering problem) as $t \to \infty$."

This definition defines a soliton by its associated eigenvalue as discussed in Sec. 3.1. It allows for a clear delineation between other hyperbolic-secant, solitary wave solutions, such as (2.36) and (2.37), and solitons, and also provides a clear foundation for what it meant by "robust to collision:" the underlying eigenvalues do not change during collision.

There is one additional modification of this definition that should be added when the soliton is considered in an engineering context, where not all parameters and conditions may be ideal. In these cases we look to first see if the ideal model for the system is indeed a soliton system as defined above. We then look to see if the non-idealities modify the behavior of the pulses in the system over the region, either in time or space, of interest. If the behavior and soliton dynamics of these pulses are preserved we call them "solitons," although in a strict mathematical sense they do not conform to the definition above. One example case is the lossy KdV discussed in Section 4.2. Over a long time period the pulse changes significantly; however, over a short time period it is quasi-static and maintains its soliton relationship of amplitude, width, and velocity. Furthermore, we will show in Chapter 6 that over short distances and times, the dynamics and quantitative predictions of inverse scattering are also maintained. Hence, in a very real sense the pulses in a lossy KdV are solitons when viewed over a short enough time or space. One can simply consider a larger time/space window as a succession of these soliton systems such that the pulse is continuously considered a soliton.

3.7 Transient Solutions of the KdV

In this chapter we have described in depth the transient solution of the KdV equation using the inverse scattering method. Inverse scattering allows us to solve for the solution at a time t given an initial profile, $u(x, t = 0)$. There is another method for calculating transient solutions known as Hirota's Direct Method. It requires that we know the solitons (amplitude/eigenvalue/velocity and phase) already, however. This might seem a pathological scenario, but the key is that we only need to specify the system at one point in time, say when all solitons are separate, and Hirota's method will allow us to calculate their time evolution. This is very helpful if one wants to analytically evaluate soliton collisions for a given number of solitons that are specified initially. From a practical standpoint, it is less useful since we must know what solitons are in the system before finding the solution.

3.7.1 Hirota's Direct Method

In 1971 Hirota introduced the first (and only) method for solving the KdV equation directly for the dynamics of soliton collisions [50]. Inspired by the recent solution of the KdV by *indirect* methods (see next section), Hirota sought a direct solution for the KdV that might also be more widely applied than the limited application of the indirect methods. We will not re-derive Hirota's work, but rather give the results of his brilliant derivation, which are surprisingly straightforward. A more detailed description is given by Hirota himself in his book [41].

This method requires a **known** set of solitons as a starting point and solves for the interaction, or collision, of these solitons over time. This method *does not* derive the solitons themselves, and hence is limited in application to those systems where we already know what solitons are present.

We start with the KdV equation,

$$u_t + 6uu_x + u_{xxx} = 0, \tag{3.60}$$

and transform it with:

$$u = 2 \cdot (\log f)_{xx}. \tag{3.61}$$

The transformed equation is then written in Hirota's bilinear form:

$$D_x(D_t + D_x^3)f \cdot f = 0, \tag{3.62}$$

where D_x is a modified form of the differential operator, $\frac{\partial}{\partial x}$, defined by Hirota himself. We then seek a solution to (3.62) by expanding f in a perturbation series:

$$f = 1 + \epsilon f_1 + \epsilon^2 f_2 + \epsilon^3 f_3 + \ldots, \tag{3.63}$$

which leads to a set of equations for each power of ϵ in terms of the lower order solutions. Just as with any perturbation solution, once we have the lowest order solution we can solve for the remaining higher orders. We write this lowest order term, f_1, in terms of the initial n solitons as follows:

$$u_n(x,t) = \frac{\kappa^2}{2} \cdot \text{sech}^2(\eta_n) \tag{3.64}$$

$$\eta_n = \kappa x - \kappa^3 t + x_o \tag{3.65}$$

$$f_1 = \exp(\eta_1) + \exp(\eta_2) + \ldots + exp(\eta_n). \tag{3.66}$$

With f_1 the higher order terms, e.g., $f_2, f_3, \ldots$, can be found. Fortunately, we find that the series truncates and an *exact* solution can be found for a finite set of terms, in fact, amazingly few terms. The following gives the perturbation series, f, for one-, two-, and three-soliton systems:

$$\text{1 Soliton: } f = 1 + \exp(\eta_1) \tag{3.67}$$

$$\text{2 Solitons: } f = 1 + \exp(\eta_1) + \exp(\eta_2) + a_{12} \cdot \exp(\eta_1 + \eta_2) \tag{3.68}$$

$$\begin{aligned}\text{3 Solitons: } f = {} & 1 + \exp(\eta_1) + \exp(\eta_2) + \exp(\eta_3) + a_{12} \cdot \exp(\eta_1 + \eta_2) \\ & + a_{13} \cdot \exp(\eta_1 + \eta_3) + a_{23} \cdot \exp(\eta_2 + \eta_3) \\ & + a_{123} \cdot \exp(\eta_1 + \eta_2 + \eta_3)\end{aligned} \tag{3.69}$$

$$a_{ij} = \left(\frac{\kappa_i - \kappa_j}{\kappa_i + \kappa_j}\right)^2, \ (i, j = 1, 2, 3) \tag{3.70}$$

The final solution, u, is then found by using the appropriate f and the transformation (3.61). This method was used to generate the solution for the two-soliton collision shown in Figs. 2.3-2.5.

The key difference between the ISM and Hirota's method is that with the ISM we started with a profile at $t = 0$ and derived the *unknown* solitons that emerged, rather than solving for the profile at $t = 0$ that occurs when two *known* solitons collide. Of course, if we use 3.54 in Hirota's method, we achieve the *exactly* same time evolution as with the inverse scattering method.

3.7.2 Transient Solution Summary

In the previous sections we looked at several different methods for examining the unique dynamics of solitons. We summarize the assumptions and benefits of these techniques as:

- **Hirota's Direct Method**: This technique allows for an analytical solution over all time of the evolution of a *known* set of solitons. This method is very useful for examining soliton dynamics, in particular soliton collisions.

- **ISM Reflectionless Potential**: This method provides the *exact* solution over all time of the evolution of an initial reflectionless profile as it evolves into a set of discrete solitons. This method tells us everything about the system, however, it is *limited* to reflectionless potentials, of which there are very few.

- **ISM Non-reflectionless Potential**: This method allows us to find the asymptotic solution for a generic profile, assuming that we can find the bounded state eigenvalues. This solution is in the form of a discrete set of independent solitons, for which we know the shape and velocity but not the phase. This method is helpful for many engineering applications because it tells us the output for a generic initial profile, even though it does not give us the detailed evolution of the transient process.

3.8 The Three Faces of the KdV Soliton

We conclude this chapter by making a note that the soliton solution of the KdV has been written in several different forms, e.g., (2.2) and (3.47). A soliton is a single parameter solution and as such we can use any single parameter to define the solution. In general there are three different parameterizations of the KdV that are used, each of which emphasizes a different perspective.

The first shows explicitly the amplitude-dependent width and velocity:

$$V(t) \quad = \quad A \cdot \mathrm{sech}^2 \left[\sqrt{\frac{A}{2}} (x - 2At) \right] . \tag{3.71}$$

The second form reformulates the soliton in terms of its velocity such that a soliton can be completely defined by its characteristic velocity. This form is seen in many mathematical texts:

$$V(t) \quad = \quad \frac{c}{2} \cdot \mathrm{sech}^2 \left[\frac{\sqrt{c}}{2} (x - ct) \right] . \tag{3.72}$$

The third form writes the soliton solution in terms of the eigenvalues of the ISM:

$$V(t) \quad = \quad 2\kappa^2 \cdot \mathrm{sech}^2 \left[\kappa (x - 4\kappa^2 t) \right] . \tag{3.73}$$

All three representations are equivalent; however, the use of one over the other tends to be confusing initially until one understands that each form simply represents a different perspective.

4

Conservative and Dissipative Soliton Systems

CONTENTS

This chapter concludes Part I with the discussion of two properties of KdV systems. The first one is a set of infinite conservation relations, e.g., conservation laws. The second is the dynamics of KdV soliton in the presence of dissipation. We group these two concepts together because the dissipation analysis leverages the conservation concept.

4.1 Conservation Laws

Conservation laws are a well known concept in science. We think of conservation of momentum, energy, mass, etc. and use these conserved quantities to solve under specified problems. To introduce the idea of conservation laws for the KdV equation, we first look at a simple conservation of mass problem where $\rho(x,t)$ is the density and $v(x,t)$ is the velocity,

$$\frac{\partial \rho}{\partial t} + \frac{\partial}{\partial x}(\rho\, v) = 0. \tag{4.1}$$

This simply states that the change in time of the density is equal to the flux into/out of an area, x. This makes physical sense, as a change in density must be due to mass flowing into or out of the system. Now let's integrate (4.1) with respect to x.

$$\frac{\partial}{\partial t}\int_{-\infty}^{\infty} \rho\, dx + [\rho v]_{-\infty}^{\infty} = 0. \tag{4.2}$$

Assuming that the flow, ρv, is constant at $|x| \to \infty$, there is no net influx of density into the system, then

$$\frac{\partial}{\partial t}\int_{-\infty}^{\infty} \rho\, dx = 0, \tag{4.3}$$

telling us that the total mass (integral of ρ) does not change with time, as one would expect.

We can generalize this as

$$\frac{\partial T}{\partial t} + \frac{\partial X}{\partial x} = 0. \tag{4.4}$$

T and X may depend on t, x, or derivatives of some function *with respect to* x. They may not depend on any derivatives with respect to t. For the KdV system, we can define Ts and Xs in terms of $x, t, u_x, u_x x, \ldots$. If X is constant at $|x| \to \infty$, we can see from the conservation of mass example that

$$\int_{-\infty}^{\infty} T\, dx = \text{constant}. \tag{4.5}$$

Let us begin with setting $T_1 = u$ and $X_1 = u_{xx} + 3u^2$ to obtain

$$\frac{\partial u}{\partial t} + \frac{\partial}{\partial x}(u_{xx} + 3u^2) = 0 \tag{4.6}$$

$$u_t + 6uu_x + u_{xxx} = 0, \tag{4.7}$$

which is just the KdV equation. This means that

$$\int_{-\infty}^{\infty} u\, dx = \text{constant}, \tag{4.8}$$

or that integral of u is a conserved quantity. For application of the KdV in water waves, this is the same as conservation of mass. We now look at two other conservation examples

$$T_2 = \frac{1}{2}u^2, \tag{4.9}$$

$$X_2 = uu_{xx} - \frac{1}{2}u_x^2 + 2u^3, \tag{4.10}$$

$$\int_{-\infty}^{\infty} u^2\, dx = \text{constant}, \tag{4.11}$$

and

$$T_3 = u^3 - \frac{1}{2}u_x^2, \tag{4.12}$$

$$X_3 = \frac{9}{4}u^4 + 3u^2u_{xx} - 6uu_x^2 - u_xu_{xxx} + \frac{1}{2}u_{xx}^2, \tag{4.13}$$

$$\int_{-\infty}^{\infty}(u^3 - \frac{1}{2}u_x^2)\,dx = \text{constant}. \tag{4.14}$$

These three correspond to conservation of mass, momentum, and energy [3], which are the three commonly known conserved quantities in physics. The KdV equation, however, does not stop there. It has many, many more conserved quantities. In fact, there are an infinite number of them [51]!! The next two Ts are[1]

$$T_4 = 5u^4 - 10uu_x^2 + u_{xx}^2, \tag{4.15}$$

$$T_5 = 21u^5 - 105u^2u_x^2 + 21uu_{xx}^2 - u_{xxx}^2. \tag{4.16}$$

We will utilize the conservation of u^2 in the next section to investigate dissipative systems.

4.2 The Lossy KdV

Up to this point in the book we have assumed a loss-less KdV. Unfortunately, most real-world systems will incur some form of loss and it is therefore important to look at the effects of loss on solitons. The lossy KdV equation was studied in 1970 by Ott and Sudan [42] for various forms of loss. One of the primary forms of loss for the NLTL is in the form $\epsilon_1 u_{xx}$, which results in the KdV-Burger's equation:

$$u_t + 6uu_x + u_{xxx} = \epsilon_1 u_{xx}. \tag{4.17}$$

[1] The coefficients of the conserved quantities listed in this section differ from [3] and [51]. This is becasue [3] uses $-6uu_x$, [51] uses $+uu_x$ and we use $+6uu_x$ for the second term of the KdV, (2.1).

We can solve this equation by looking at how the total energy of the system (conserved quantity from the previous section),

$$E = \int_{-\infty}^{\infty} u^2 \, dx, \tag{4.18}$$

changes with time and then examine the effects of that change on the soliton solution [19,42]. In order to write the equation in terms of energy, we multiply both sides by u and rearrange:

$$\frac{1}{2}(u^2)_t = (-6u^2 u_x - u u_{xxx}) + \epsilon_1 u u_{xx}, \tag{4.19}$$

and then integrate over x:

$$\frac{\partial}{\partial t} \int_{-\infty}^{\infty} \frac{1}{2} u^2 \, dx = \int_{-\infty}^{\infty} (-6u^2 u_x - u u_{xxx}) \, dx + \int_{-\infty}^{\infty} \epsilon_1 u u_{xx} \, dx, \tag{4.20}$$

or,

$$\begin{aligned} \frac{\partial}{\partial t} \int_{-\infty}^{\infty} \frac{1}{2} u^2 \, dx &= \int_{-\infty}^{\infty} -(2u^3 + u_{xx}u - \frac{1}{2}(u_x)^2)_x \, dx + \\ &\quad \int_{-\infty}^{\infty} \epsilon_1 u u_{xx} \, dx \qquad (4.21) \\ &= -\left[2u^3 + u_{xx}u - \frac{1}{2}(u_x)^2\right]_{-\infty}^{\infty} + \\ &\quad \int_{-\infty}^{\infty} \epsilon_1 u u_{xx} \, dx, \qquad (4.22) \end{aligned}$$

where we have written the time derivative explicitly. The first term on the RHS of (4.22) is equal to zero, since for a soliton, both u and all of its derivatives tend to zero as $x \to \infty$. This allows us to simplify (4.22) to:

$$\frac{\partial}{\partial t} \int_{-\infty}^{\infty} u^2 \, dx = \int_{-\infty}^{\infty} 2\epsilon_1 u u_{xx} \, dx, \tag{4.23}$$

which tells us how the energy of the soliton changes with time.

Assuming a soliton of the form,

$$u(x,t) = A(t) \cdot \text{sech}^2 \left[\sqrt{\frac{A(t)}{2}} \left(x - 2\int_0^t A(\tau)\, d\tau \right) \right], \quad (4.24)$$

we can write the LHS of (4.23) as:

$$\frac{\partial}{\partial t} \int_{-\infty}^{\infty} u^2\, dx = \frac{\partial}{\partial t} \frac{4\sqrt{2}}{3} A^{\frac{3}{2}}; \quad (4.25)$$

and the RHS of (4.23) as:

$$\int_{-\infty}^{\infty} 2\epsilon_1 u u_{xx}\, dx = -\epsilon_1 \frac{16\sqrt{2}}{15} A^{\frac{5}{2}}. \quad (4.26)$$

Combining these two results we find the evolution of the soliton amplitude is given by:

$$\frac{\partial}{\partial t} A = -\frac{8}{15} A^2, \quad (4.27)$$

or,

$$A(t) = \frac{A(0)}{1 + \frac{8}{15} A(0) \epsilon_1 t}. \quad (4.28)$$

The result is that a lossy soliton slowly reduces in amplitude as it propagates according to (4.28) [note that the $\frac{8}{15}$ in the denominator of (4.28) differs from the solution in [42], which gives $\frac{16}{15}$; this difference may be accounted for in footnote (10) of [42] which appears to be incorrect]. This analysis was based upon the assumption that the soliton maintained its relationship between amplitude and width/velocity, i.e., (4.24). As a result, a lossy soliton will widen as it reduces in amplitude. It will also reduce in velocity, causing the difference in position between a loss-less soliton and a lossy soliton to grow with time. This is accounted for by integrating the change in velocity over the time interval $t = 0$ to $t = t'$. The resulting damped soliton is then given as:

$$u(x,t') = A(t') \cdot \text{sech}^2 \left[\sqrt{\frac{A(t')}{2}} \left(x - 2\int_0^{t'} A(\tau)\, d\tau \right) \right]. \quad (4.29)$$

The instantaneous solution at time, t, is given as:

$$u(x,t) = A(t) \cdot \text{sech}^2 \left[\sqrt{\frac{A(t)}{2}} (x - 2A(t) \cdot t) \right]. \quad (4.30)$$

The result is that for small time intervals the lossy soliton appears as any normal soliton. Over long time intervals, it will reduce in amplitude and velocity and increase in width.

With this general method from Ott [42], we can also solve (4.17) for loss terms other than u_{xx}. For a loss term of the form $\epsilon_2 u$, Ott has shown that (4.28) is simply rewritten as:

$$A(t) = e^{\left(\frac{-4\epsilon_2 t}{3}\right)}. \quad (4.31)$$

The damping soliton solution from $\epsilon_2 u$ is given by (4.29) and (4.30) with $A(t)$ according to (4.31). This method can be applied to other loss terms, however, $\epsilon_1 u_{xx}$ and $\epsilon_2 u$ are the two main loss mechanism electrical solitons. Figure 4.1 shows (4.31) and (4.28), respectively, plotted against numerical simulation of the KdV equation. They show good agreement between theory and simulation.

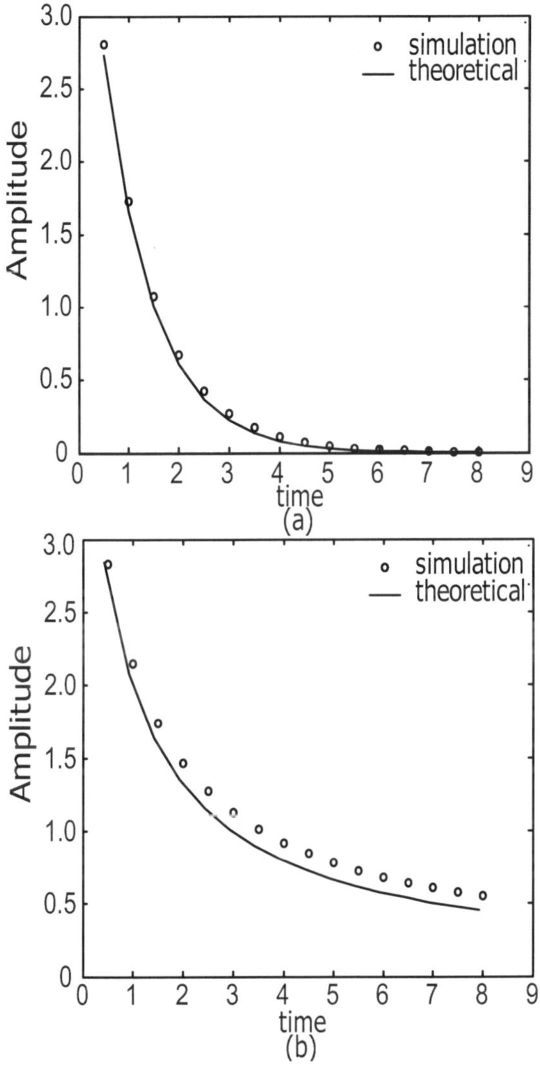

FIGURE 4.1
Loss mechanisms in KdV: (a) Frequency-independent loss, ε_2; (b) Frequency-dependent loss, ε_1.

Part II

Electrical Solitons: Design

5

Electrical Nonlinear Transmission Line and Electrical Solitons

CONTENTS

In Chapter 2 we reviewed the basic properties and dynamics of the KdV soliton. These same properties and dynamics are found in many other physical systems in nature such as acoustic waves in plasma and vibrations in anharmonic lattices (spring-mass). In the electrical domain, a structure known as the nonlinear transmission line (NLTL) also supports KdV solitons in the form of voltage waves. This chapter will review the NLTL structure and the electrical solitons it supports, including the link with the anharmonic lattice and the KdV equation.

5.1 The Nonlinear Transmission Line

The nonlinear transmission line (NLTL) can be constructed from a linear transmission line, such as a coplanar stripline (CPS), by periodically loading it with varactors, such as reverse-biased *pn* junction diodes or MOS capacitors [11] [See Fig. 5.1(a)]. Varactors are nonlinear capacitors, whose capacitance changes with the applied voltage. An NLTL can be alternatively obtained by replacing linear capacitors of an artificial LC transmission line with varactors as shown in Fig. 5.1(b). This latter implementation is an exact analog to the nonlinear lattice studied by Fermi, Pasta and Ulam, where the masses are replaced by inductors and the nonlinear springs by varactors.

The NLTL is a nonlinear dispersive system which supports the forma-

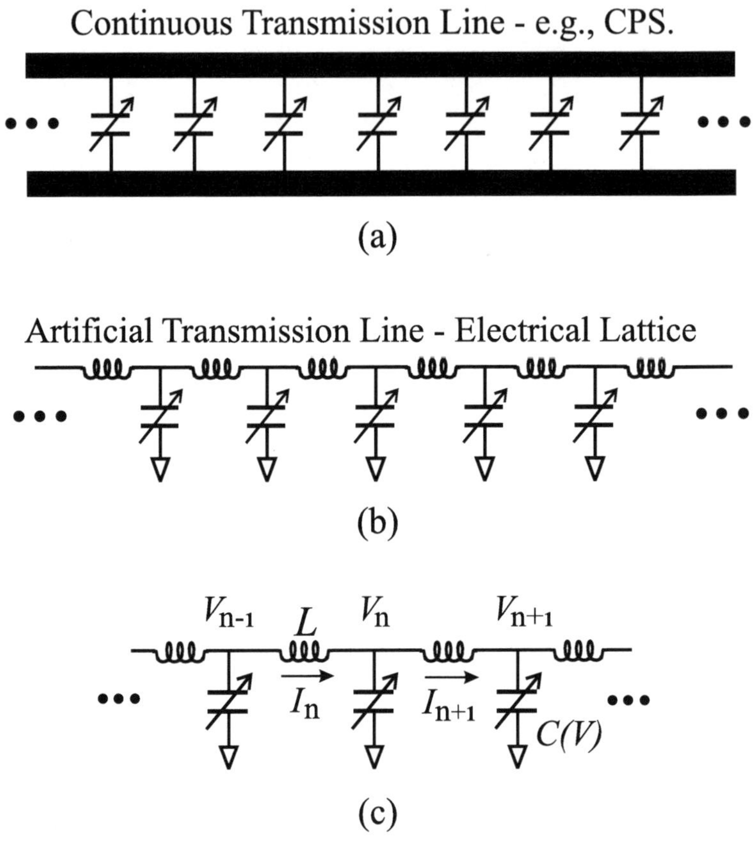

FIGURE 5.1
(a) Continuous NLTL. (b) Artificial NLTL - Electrical Nonlinear Lattice. (c) Nodal voltages and currents for NLTL.

tion and propagation of electrical solitons. The nonlinearity originates from the varactors. The dispersion arises from the structural periodicity.

5.2 Toda Lattice

We can derive the underlying dynamics of the NLTL by solving the nodal equations for the voltages and currents shown in Fig. 5.1(c):

$$\begin{aligned} V_{n-1} - V_n &= L\frac{\partial I_n}{\partial t} \\ V_n - V_{n+1} &= L\frac{\partial I_{n+1}}{\partial t} \\ I_n - I_{n+1} &= \frac{\partial Q_n(V)}{\partial t}. \end{aligned} \tag{5.1}$$

Combining these yields,

$$\frac{\partial^2 Q_n(V)}{\partial t^2} = \frac{1}{L}(V_{n+1} + V_{n-1} - 2V_n), \tag{5.2}$$

where $Q_n(V)$ is the charge on the nth capacitor.

If we choose $Q_n(V)$ to be:

$$Q_n(V) = Q_0 \ln\left[1 + \frac{V_n}{F_0}\right], \tag{5.3}$$

the system can be rewritten as:

$$Q_0 L\frac{\partial^2}{\partial t^2}\ln\left[1 + \frac{V_n}{F_0}\right] = (V_{n+1} + V_{n-1} - 2V_n), \tag{5.4}$$

which is nothing more than a special lattice known as the Toda lattice, where the force between masses is replaced by the voltage between sections of the NLTL [5] (F_0 is simply a constant). Since the Toda lattice has been shown to be a perfect soliton system, the NLTL with the charge-voltage relationship (5.3) is an exact soliton system.[1]

[1] The Toda lattice does differ from the KdV, however, both are exact soliton systems. In the small amplitude limit the KdV is a good approximation of the Toda lattice.

This places the NLTL on a firm foundation as a soliton medium. The Toda lattice, however, is very cumbersome to work with since most of the computations are done with large matrices or with elliptic functions, which are unfamiliar to most and difficult to manipulate. For this reason it is advantageous to simplify (5.3) and make several other approximations to achieve a more manageable form for the NLTL.

5.3 NLTL Lattice

We begin by assuming a simpler form for the charge-voltage relationship of the varactors in the NLTL, which for small amplitudes is a good approximation of (5.3):

$$\ln\left[1+\frac{V_n}{F_0}\right] \approx \frac{V_n}{F_0} - \frac{V_n^2}{2F_0^2}. \tag{5.5}$$

This allows us to rewrite (5.4) as:

$$\frac{Q_0}{F_0}L\frac{\partial^2}{\partial t^2}\left(1-\frac{V_n}{2F_0}\right)V_n = (V_{n+1}+V_{n-1}-2V_n). \tag{5.6}$$

The LHS is in terms of charge on the capacitor which is related to the voltage-dependent capacitance by:

$$\begin{aligned} C(V) &= \frac{\partial Q(V)}{\partial V} &(5.7)\\ &= C_0(1-2bV), &(5.8)\end{aligned}$$

with $C_0 = Q_0/F_0$ and $2b = 1/F_0$.

It is important to note that the units of b are [1/Volt] and that the quantity, bA, where A is the amplitude of the soliton, is therefore dimensionless. Thus, we can consider bA as the **normalized amplitude**. Its physical meaning is simple: bA represents the amount of capacitance reduction for a given A, or bA is the degree of the nonlinearity for a given amplitude. The dynamics of the NLTL and solitons are based upon the nonlinearity, which is determined by bA, and not the absolute values of b and A, as we will see at the end of this section.

With (5.7), the final result is:

$$C_0 L \frac{\partial}{\partial t}\left[(1-2bV_n)\frac{\partial V_n}{\partial t}\right] = (V_{n+1}+V_{n-1}-2V_n). \tag{5.9}$$

This equation as it stands does not have an analytical solution [5]. We therefore make a *continuum* approximation where we assume that the wavelength of the signal of interest is long enough that we can treat the discrete lattice as a continuous system. We start by expanding the nodal voltages in a Taylor expansion:

$$V_{n\pm1} = V_n \pm \frac{\partial V_n}{\partial n} + \frac{1}{2!}\frac{\partial^2 V_n}{\partial n^2} \pm \frac{1}{3!}\frac{\partial^3 V_n}{\partial n^3} + \frac{1}{4!}\frac{\partial^4 V_n}{\partial n^4} \pm \ldots \tag{5.10}$$

and then write (5.9) using $\frac{\partial V}{\partial t} - V_t$ and $VV_t = \frac{1}{2}(V^2)_t$:

$$V_{tt} - b(V^2)_{tt} = \frac{1}{LC_0}V_{nn} + \frac{1}{12LC_0}V_{nnnn}. \tag{5.11}$$

We now follow the same process as in Sec. 2.2 by introducing $\xi = n - ct$:

$$\left(c^2 - \frac{1}{LC_0}\right)V_{\xi\xi} = bc^2(V^2)_{\xi\xi} + \frac{1}{12LC_0}V_{\xi\xi\xi\xi}. \tag{5.12}$$

We now integrate once with respect to ξ,

$$\left(c^2 - \frac{1}{LC_0}\right)V_{\xi} = bc^2(V^2)_{\xi} + \frac{1}{12LC_0}V_{\xi\xi\xi} + C_1, \tag{5.13}$$

and then integrate a second time,

$$\left(c^2 - \frac{1}{LC_0}\right)V = bc^2(V^2) + \frac{1}{12LC_0}V_{\xi\xi} + C_1 V + C_2. \tag{5.14}$$

We now multiply by V_ξ on both sides and note that $(V^2)_\xi = 2VV_\xi$, $(V^3)_\xi = 3V^2V_\xi$ and $(V_\xi^2)_\xi = 2V_\xi V_{\xi\xi}$,

$$\begin{aligned}\left(c^2-\frac{1}{LC_0}\right)\frac{1}{2}(V^2)_\xi &= \frac{bc^2}{3}(V^3)_\xi+\frac{1}{24LC_0}(V_\xi^2)_\xi\\ &+\frac{C_1}{2}(V^2)_\xi+C_2V_\xi. \end{aligned} \tag{5.15}$$

Integrating again, we obtain an equation similar to (2.13):

$$\begin{aligned} V_\xi^2 &= -8LC_0bc^2(V^3)+12(LC_0c^2-1-LC_0C_1)V^2\\ &-12LC_0C_2V-12LC_0C_3. \end{aligned} \tag{5.16}$$

Instead of finding the solitary wave solution from the periodic solution as we did in Sec. 2.2.1, we note that for the solitary wave solution the voltage, V, and all of its derivatives are zero as $\xi \to \infty$. Therefore C_1 and C_2 are both zero, yielding:

$$V_\xi^2 = -8LC_0bc^2(V^3)+12(LC_0c^2-1)V^2. \tag{5.17}$$

Using $v_0 = \frac{1}{\sqrt{LC_0}}$ and $V_\xi = \frac{\partial V}{\partial \xi}$ this can be rewritten as:

$$\frac{\partial V}{\partial \xi} = 2\sqrt{\frac{3(c^2-v_0^2)}{v_0^2}}\cdot V\cdot\sqrt{1-\frac{2bc^2}{3(c^2-v_0^2)}\cdot V} \tag{5.18}$$

and then integrated,

$$\int_{V_0}^{V}\frac{dV}{2\sqrt{\frac{3(c^2-v_0^2)}{v_0^2}}\cdot V\cdot\sqrt{1-\frac{2bc^2}{3(c^2-v_0^2)}\cdot V}} = \int_{\xi_0}^{\xi} d\xi. \tag{5.19}$$

Using

$$X^2 = \frac{2bc^2}{3(c^2-v_0^2)}\cdot V \tag{5.20}$$

$$dV = \frac{3(c^2-v_0^2)}{bc^2}\cdot X\cdot dX \tag{5.21}$$

and noting that,

$$\int \frac{dX}{X\sqrt{1-X^2}} = \text{arcsech(X)}, \tag{5.22}$$

the solitary wave solution is given as:

$$V(n,t) = \frac{3(c^2 - v_0^2)}{2bc^2}\text{sech}^2\left[\frac{\sqrt{3(c^2 - v_0^2)}}{v_0}(n - ct)\right], \tag{5.23}$$

or,

$$V(n,t) = A \cdot \text{sech}^2\left[\sqrt{\frac{6bA}{3-2bA}}\left(n - v_0\sqrt{\frac{3}{3-2bA}}t\right)\right]. \tag{5.24}$$

We conclude this section by noting that the argument of sech in (5.24) is written in terms of the normalized amplitude, bA, such that the soliton properties, e.g., velocity and width, are determined by this parameter.

5.4 KdV Approximation of the NLTL

The solitary waves on the nonlinear transmission lines, (5.24), are very similar to the KdV solitons studied in Chapter 3. In fact, (5.11) can be reduced to the KdV equation by using a technique called the reductive perturbation method [5, 52] . This is an important fact, as it allows us to use the properties and dynamics described in Chapter 3 to describe the dynamics of the NLTL (for small amplitudes).

The reductive perturbation method begins with transforming our system from the lab reference frame to a reference frame that moves at a nominal velocity, in our case, a reference frame that moves at the fixed velocity $v_0 = 1/\sqrt{LC_0}$; note that this is different than $\xi = n - ct$, which moves at the amplitude-dependent velocity, c, of the soliton. In addition to changing reference frames, we scale both the reference frame and time by ε as follows:

$$\begin{aligned} S &= \varepsilon^{1/2}(n - v_0 t) \\ T &= \varepsilon^{3/2} t \end{aligned}$$

and note that the derivatives with respect to n and t are now:

$$\begin{aligned}\frac{\partial}{\partial n} &= \varepsilon^{1/2}\frac{\partial}{\partial S}\\ \frac{\partial}{\partial t} &= -\varepsilon^{1/2}v_0\frac{\partial}{\partial S}+\varepsilon^{3/2}\frac{\partial}{\partial T}\end{aligned}$$

We now rewrite (5.11) using $V_{NLTL} = \varepsilon V_{KdVa}$, where V_{KdVa} is the voltage in the KdV approximation[2], and the scaled coordinates,

$$\begin{aligned}\frac{1}{12}\varepsilon^3 V_{SSSS} &= -2\varepsilon^3 v_o V_{ST}+\varepsilon^4 v_o^2 V_{TT}-\varepsilon^3 b(V^2)_{SS}\\ &\quad +2\varepsilon^4 v_o(V^2)_{TS}+\varepsilon^5 V_{TT}\end{aligned} \tag{5.25}$$

Keeping terms O(ε^3) we obtain (note the ε^3 then cancel):

$$V_{ST}+\frac{b}{2\sqrt{LC_0}}(V^2)_{SS}+\frac{1}{24\sqrt{LC_0}}V_{SSSS}=0 \tag{5.26}$$

Integrating once and using $(V^2)_S = 2VV_S$, we obtain the KdV equation (with modified coefficients):

$$V_T+\frac{b}{\sqrt{LC_0}}VV_S+\frac{1}{24\sqrt{LC_0}}V_{SSS}=0. \tag{5.27}$$

The soliton solution for the KdV with these coefficients is:

$$V(t)_{KdVa} = \frac{\chi}{8b}\cdot \text{sech}^2\left[\frac{\sqrt{\chi}}{2}\left(S-\frac{\chi}{24}v_0 T\right)\right], \tag{5.28}$$

where χ is a parameter of the soliton that determines the amplitude, width and velocity (recall the soliton is defined by a single parameter). Equations (5.23) and (5.24) used amplitude, A, or velocity, c, respectively, as their parameters. Here, we simply define $\chi = 8bA$ as the parameter. Now if we scale the voltage, space and time such that $u = 4b\cdot V$, $x = S$ and $\tau = \frac{c_{KdVa}}{24}T$ we obtain

[2] We drop $KdVa$ subscript in the following to simplify writing the equations.

$$u(\tau) \;=\; \frac{c}{2}\cdot \text{sech}^2\left[\frac{\sqrt{c}}{2}\left(S - c\tau\right)\right], \tag{5.29}$$

which is exactly the same as (2.8), with the KdV velocity, c, and hence, we have shown the equivalence to the canonical KdV solution.

It is often helpful to convert (5.28) back to laboratory coordinates for comparing measurements to KdV theory. To do this we scale 5.28 using:

$$\begin{aligned} S &= \varepsilon^{1/2}(n - v_0 t) \\ T &= \varepsilon^{3/2} t \\ V_{NLTL} &= \varepsilon V_{KdVa}. \end{aligned} \tag{5.30}$$

This results in

$$\begin{aligned} V_{NLTL}(t) &= \varepsilon V_{KdVa} \\ &= \tfrac{\chi}{8b}\cdot \text{sech}^2\left[\tfrac{\sqrt{\chi}}{2}\left(n - \left(1 + \tfrac{\chi}{24}\right) v_0 t\right)\right], \end{aligned} \tag{5.31}$$

where the velocity is now $(1 + \chi/24)v_0$.

If we now compare this to (5.24), rewritten below,

$$V(n,t) \;=\; A\cdot \text{sech}^2\left[\sqrt{\frac{6bA}{3-2bA}}\left(n - v_0\sqrt{\frac{3}{3-2bA}}t\right)\right], \tag{5.32}$$

we see that there is a difference between the laboratory version of the KdV approximation and our solution to the NLTL lattice. For small amplitudes, however, we can see that they are equivalent. If we set $A = \chi/8b$ we can rewrite (5.31) as

$$V(n,t) = A\cdot \text{sech}^2\left[\sqrt{2bA}\left(n - v_0\left(1 + \frac{bA}{3}\right)t\right)\right]. \tag{5.33}$$

We now compare the coefficients in (5.33) and (5.32), by looking at their ratio, denoted β (ideally $\beta = 1$ if they are equivalent):

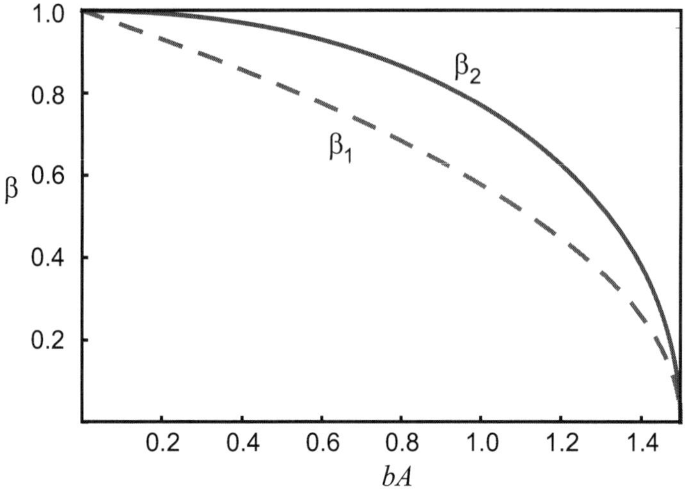

FIGURE 5.2
Ratios, β_1 and β_2, of the coefficients of the NLTL lattice and KdV approximation versus bA, normalized amplitude.

$$\beta_1 = \frac{\sqrt{2bA}}{\sqrt{\frac{6bA}{3-2bA}}}$$

$$\beta_2 = \frac{\frac{bA}{3}}{\sqrt{\frac{3}{3-2bA}}}. \tag{5.34}$$

Figure 5.2 show the ratios versus the normalized amplitude, bA. For $bA < 0.3$ the ratios are within 10%. Typical b for nonlinear capacitors are 0.05 to 0.2, which corresponds to amplitudes of less than 6V to 1.5V, respectively, which are well within the operating range of many practical applications.

5.5 The Lossy NLTL

In Chapter 2 we examined the lossy-KdV system and showed that in such a system a soliton will propagate with a time reducing amplitude.

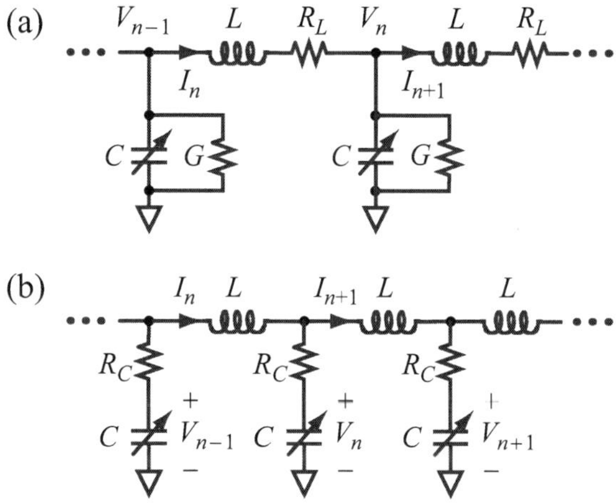

FIGURE 5.3
Loss mechanisms in NLTL: (a) Frequency-independent loss due to serial loss of the inductors and parallel loss of the capacitors; (b) Frequency-dependent loss due to the series resistance of the varactors.

The soliton's width and velocity scale accordingly with this reducing amplitude. In the NLTL we use these results to model the lossy NLTL. The one difference is that we use (5.24) instead of (4.24), so that the instantaneous NLTL soliton is:

$$V(n,t) = A(t) \cdot \text{sech}^2 \left[\sqrt{\frac{6bA(t)}{3-2bA(t)}} \left(n - v_0 \sqrt{\frac{3}{3-2bA(t)}} t \right) \right].$$

In the NLTL there are two major forms of loss. The first is frequency-*independent* loss, which, for example, is caused by series resistance in the inductor and resistance in *parallel* with the capacitor, Fig. 5.3(a). We can use the same nodal analysis of Sec. 5.2 and the reductive perturbation method outlined in Sec. 5.4 to find the damping term.

We begin with the nodal equations for Fig. 5.3(a),

$$\begin{aligned} \frac{\partial}{\partial t} Q_n + GV_n &= I_n - I_{n+1} \\ L\frac{\partial}{\partial t} I_n + RI_n &= V_{n-1} - V_n. \end{aligned} \tag{5.35}$$

Combining as we did for the loss-less NLTL, we obtain

$$L\frac{\partial^2}{\partial t^2}Q_n + R\frac{\partial}{\partial t}Q_n + LG\frac{\partial}{\partial t}V_n + RGV_n = V_{n+1} + V_{n-1} - 2V_n. \quad (5.36)$$

Then, we apply the charge-voltage relation of the varactor

$$Q(V_n) = C_0(V_n - bV_n^2), \quad (5.37)$$

and expanding the voltage in a Taylor series

$$\begin{aligned} V(n \pm 1) &= V(n) \pm \frac{\partial V(n)}{\partial n} + \frac{1}{2!}\frac{\partial^2 V(n)}{\partial n^2} \pm \\ &\frac{1}{3!}\frac{\partial^3 V(n)}{\partial n^3} + \frac{1}{4!}\frac{\partial^4 V(n)}{\partial n^4} \end{aligned} \quad (5.38)$$

Keeping derivatives O(4), we obtain

$$V_{n+1} + V_{n-1} - 2V_n = V_{nn} + \frac{1}{12}V_{nnnn}. \quad (5.39)$$

Substituting into (5.36) we obtain

$$\begin{aligned} v_o^2\left(\frac{\partial^2 V}{\partial n^2} + \frac{1}{12}\frac{\partial^4 V}{\partial n^4}\right) &= \frac{\partial^2}{\partial t^2}\left(V - \frac{V^2}{b}\right) + \frac{R}{L}\frac{\partial}{\partial t}\left(V - \frac{V^2}{b}\right) + \\ &\frac{G}{C_d}\frac{\partial V}{\partial t} + \frac{RG}{LC_d}V \end{aligned} \quad (5.40)$$

We then utilize the reductive perturbation method [53], with $V \to \varepsilon V$, $R/L \to \varepsilon^{3/2}(R/L)$, $G/C_o \to \varepsilon^{3/2}(G/C_o)$ and a scaled moving frame S and time T

$$\begin{cases} S = \varepsilon^{1/2}(n - v_0 t) \\ T = \varepsilon^{3/2}t \end{cases} \Rightarrow \begin{cases} \dfrac{\partial}{\partial n} = \varepsilon^{1/2}\dfrac{\partial}{\partial S} \\ \dfrac{\partial}{\partial t} = -\varepsilon^{1/2}v_0\dfrac{\partial}{\partial S} + \varepsilon^{3/2}\dfrac{\partial}{\partial T}. \end{cases} \quad (5.41)$$

Keeping terms up to $O(\varepsilon^3)$, reduces (5.40) into a simpler form

$$V_T + bv_0 V V_S + \frac{v_0}{24}V_{SSS} = \frac{1}{2}\left(\frac{R}{L} + \frac{G}{C}\right)V, \quad (5.42)$$

where $v_0 = 1/\sqrt{LC_0}$.

The RHS corresponds to ε_2 in Sec. 4.2 for frequency-independent loss:

$$\epsilon_2 V = \frac{1}{2} \cdot (\frac{R_L}{L} + \frac{G}{C}) \cdot V \tag{5.43}$$

and the resulting amplitude reduction versus time using (4.31) is:

$$A(t) \quad = \quad A(0) \cdot e^{\left[-\frac{2}{3}\left(\frac{R_L}{L}+\frac{G}{C}\right)\cdot t\right]}. \tag{5.44}$$

The second form of loss is frequency-*dependent* loss, which, for example, is caused by the resistance in *series* with the capacitor, Fig. 5.3(b). Using the same techniques as above, we can determine the frequency-dependent loss equation.

We write the nodal equations as:

$$\begin{aligned} I_n - I_{n+1} &= \frac{\partial Q(V_n)}{\partial t}, \\ L\frac{\partial I_n}{\partial t} &= \left[V_{n-1} + R_C\frac{\partial Q(V_{n-1})}{\partial t}\right] - \left[V_n + R_C\frac{\partial Q(V_n)}{\partial t}\right], \\ L\frac{\partial I_{n+1}}{\partial t} &= \left[V_n + R_C\frac{\partial Q(V_n)}{\partial t}\right] - \left[V_{n+1} + R_C\frac{\partial Q(V_{n+1})}{\partial t}\right]. \end{aligned} \tag{5.45}$$

Note that, for convenience, we have chosen V_n as the voltage on the nonlinear capacitor. Combining these yields:

$$L\frac{\partial^2 Q(V_n)}{\partial t^2} = (V_{n+1}+V_{n-1}-2V_n)+R_C\frac{\partial}{\partial t}[Q(V_{n+1})+Q(V_{n-1})-2Q(V_n)]. \tag{5.46}$$

Then, we apply the charge-voltage relation of the varactor

$$Q(V_n) = C_0(V_n - bV_n^2), \tag{5.47}$$

and writing the voltage as a Taylor series as above and keeping derivatives O(4) yields

$$V_{n+1} + V_{n-1} - 2V_n = V_{nn} + \frac{1}{12}V_{nnnn} \tag{5.48}$$

resulting in

$$V_{tt} - b(V^2)_{tt} = v_0^2 \left(V_{nn} + \frac{1}{12} V_{nnnn} \right) + \frac{R_C}{L} V_{nnt}, \tag{5.49}$$

where $v_0 = 1/\sqrt{LC_0}$. Note that, in the last term, we have kept only the dominant effect of dissipation. Finally, we utilize the reductive perturbation method [54], with $V \to \varepsilon V$, $R_C/L \to \varepsilon^{1/2}(R_C/L)$, and a scaled moving frame S and time T

$$\begin{cases} S = \varepsilon^{1/2}(n - v_0 t) \\ T = \varepsilon^{3/2} t \end{cases} \Rightarrow \begin{cases} \dfrac{\partial}{\partial n} = \varepsilon^{1/2} \dfrac{\partial}{\partial S} \\ \dfrac{\partial}{\partial t} = -\varepsilon^{1/2} v_0 \dfrac{\partial}{\partial S} + \varepsilon^{3/2} \dfrac{\partial}{\partial T}. \end{cases} \tag{5.50}$$

Keeping terms up to $O(\varepsilon^3)$, reduces (5.49) into a simpler form

$$V_T + b v_0 V V_S + \frac{v_0}{24} V_{SSS} = \frac{R_C}{2L} V_{SS}. \tag{5.51}$$

This corresponds to the RHS of (4.17) for frequency-dependent loss in Sec. 4.2:

$$\epsilon_1 V_{ss} = \frac{R_C}{2L} \cdot V_{ss}, \tag{5.52}$$

and the resulting amplitude reduction versus time using (4.28) is:

$$A(t) = \frac{A(0)}{1 + \frac{32}{15} A(0) \frac{b R_c}{L} \cdot t}. \tag{5.53}$$

6

The Electrical Soliton in the Lab

Michael W. Chen
Carnegie Mellon University

En Shi
Carnegie Mellon University

CONTENTS

In this chapter we demonstrate experimentally the properties of the NLTL and the electrical soliton. Through this, we achieve two goals:

1. We demonstrate that the pulses in the NLTL are indeed solitons.
2. We demonstrate the explicit link between KdV soliton theory and the NLTL lattice.

This chapter is divided into six sections which detail the practical implementation of the KdV and NLTL soliton theory. The first reviews the three solutions for the NLTL and KdV systems. The second section reviews the transformations between the theory presented in Chapters 2–5 and the experimental measurements. The third section characterizes the NLTL used in our experiments, in particular the component parameters: C_0, L, b, as well as R_L and R_C. The fourth demonstrates the power of the inverse scattering method for determining the sharpening of a pulse. The fifth section shows the damping of electrical solitons on the NLTL. The final section describes inaccuracies of simulations of the NLTL using soliton damping as an example case.

6.1 Toda Lattice, NLTL Lattice and KdV Solitons

In this section we briefly compare and contrast the soliton equations described in the previous chapters: the Toda lattice, (5.4), the NLTL lattice,(5.11), the KdV approximation of the NLTL, (5.27) and the KdV equation, (2.1). We rewrite these equations below for comparison:

$$\begin{array}{lrl} \text{Toda Lattice:} & Q_0 L\frac{\partial^2}{\partial t^2}\ln[1+\frac{V_n}{F_0}] - (V_{n+1}+V_{n-1}-2V_n) & = 0 \\ \text{NLTL Lattice:} & V_{tt} - b(V^2)_{tt} - \frac{1}{LC_0}V_{nn} - \frac{1}{12LC_0}V_{nnnn} & = 0 \\ \text{KdV-Approx.:} & V_T + \frac{b}{\sqrt{LC_0}}VV_S + \frac{1}{24\sqrt{LC_0}}V_{SSS} & = 0 \\ \text{KdV:} & u_\tau + 6uu_x + u_{xxx} & = 0 \end{array}$$

Their soliton solutions are (respectively):

$$V(n,t) = A\cdot \text{sech}^2\left[\Gamma\left(n - v_0\frac{\sqrt{A}}{\sqrt{F_0}\cdot\Gamma}t\right)\right] \tag{6.1}$$

$$V(n,t) = A\cdot \text{sech}^2\left[\sqrt{\frac{6bA}{3-2bA}}\left(n - v_0\sqrt{\frac{3}{3-2bA}}t\right)\right] \tag{6.2}$$

$$V(S,T) = A\cdot \text{sech}^2\left[\sqrt{2bA}\left(S - v_0\frac{bA}{3}T\right)\right] \tag{6.3}$$

$$u(x,\tau) = A\cdot \text{sech}^2\left[\sqrt{\frac{A}{2}}(x-2At)\right], \tag{6.4}$$

where we have rewritten (5.31) in terms of amplitude, A, $\Gamma = \text{arcsinh}\sqrt{\text{A}/\text{F}_0}$ and we have introduced the soliton solution for the Toda Lattice, (6.1) (derivation of this solution from (5.4) may be found in [5] or [55]).

Qualitatively it appears that the only common property is their sech^2 shape. Their velocities and pulse widths are, however, also quite similar, where we define pulse width, w, and velocity, c, as:

$$\begin{aligned} V(x,t) &= \text{sech}^2(B(x-Ct)) \\ w_x &= \frac{1.76}{B} \\ w_t &= \frac{1.76}{BC} \\ c &= C \end{aligned}$$

The 1.76 term is to convert the sech^2 pulse width into Full Width Half Maximum (FWHM).

Figure 6.1 shows the pulse width and normalized velocities (c/v_0) versus the normalized amplitude, bA. At small amplitudes, $bA < 0.3$, they are virtually identical. At larger amplitudes, the difference in their behavior is evident. The NLTL lattice tends to zero width and infinite velocity at a finite amplitude due to the $(3-2bA)$ term. This is simply a result of the fact that (5.5) becomes negative for large amplitudes, which is not physical; it is just a result of our approximation.

6.2 Scaling and Transformations: Lab → NLTL → KdV

In the previous chapter we have noted several transformations and scalings between the KdV theory of Chapter 2, the NLTL and the KdV approximation of the NLTL. In addition there is a transformation between the laboratory measurements, which are in time, and the spatial properties discussed previously, such as pulse width, which are in space. We summarize all of these in this section to provide a concise reference for the reader.

We begin by noting that laboratory measurements are in the time domain; the oscilloscope measures the signal in time at one spatial location. There are two effects we should note: 1) the dependance of spatial width on velocity and 2) the apparent "inversion" of waveforms in the lab. Figure 6.2 (a) shows the effect of waveform velocity on temporal width; the higher the velocity, the narrower the pulse for a given spatial width. Figure 6.2 (b) shows the apparent "inversion" of waveforms by the oscilloscope. Since the oscilloscope measures at a single spatial location and records the waveform as it appears at the measurement point, a waveform with a higher amplitude on the right, will appear inverted, with the higher amplitude on the left, in the oscilloscope. While standard for lab measurements, this point becomes important when interpreting soliton dynamics, such as amplitude dependent velocity. We will see this inversion when we examine the soliton oscillator in Chapter 9.

The dependance of temporal pulse width on velocity can be expressed as:

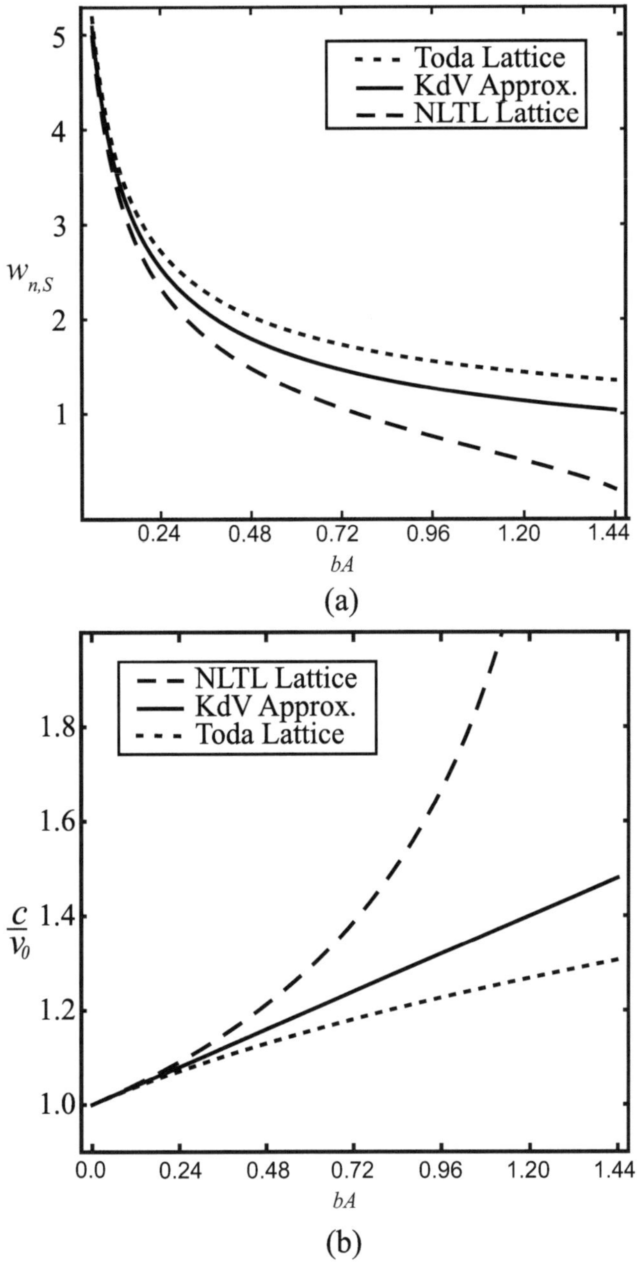

FIGURE 6.1
(a) Soliton spatial width, w, versus normalized amplitude, bA. (b) Soliton normalized velocity, c/v_0, versus normalized amplitude, bA. Note: x-axis is divided into $2b = 0.24$ segments, which correspond to $A = 2V$ segments.

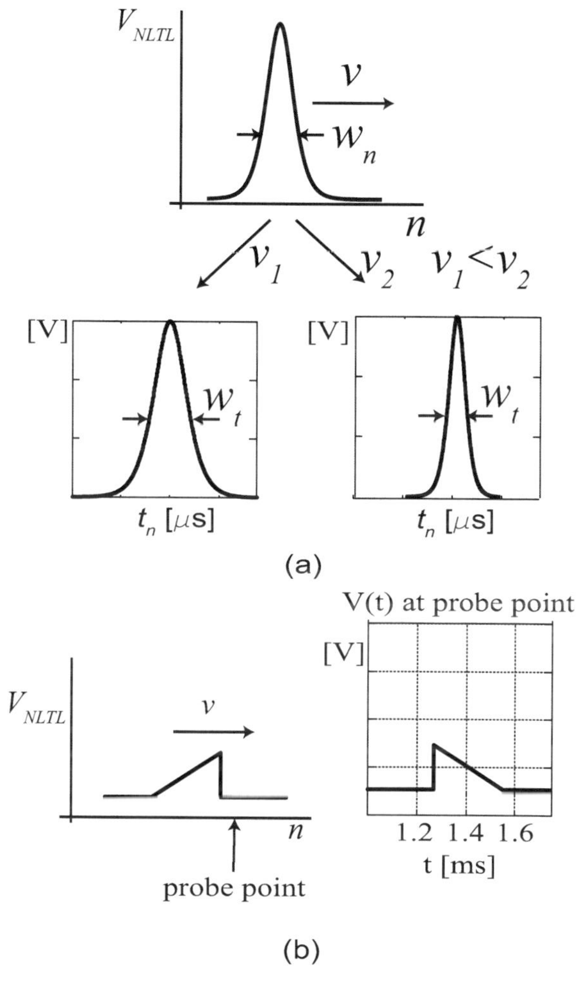

FIGURE 6.2
(a) Temporal pulse width is a function of pulse velocity. A pulse with a fixed spatial width has a narrower temporal pulse width with a higher velocity. (b) The oscilloscope creates an apparent "inversion" of the waveform since it measures the waveform at a fixed spatial location.

$$w_{time} = w_{space} \cdot \frac{1}{c}. \tag{6.5}$$

Note that the pulse width can be in *sections* and the velocity, c, in sections/second or in *meters* and *meters/second*, respectively. For the NLTL soliton the pulse width in time is given as:

$$w_{time} = \frac{3 - 2bA}{\sqrt{6bA}} \cdot \sqrt{LC_0}, \tag{6.6}$$

which, once again, is defined as Full Width Half Maximum (FWHM). In terms of l which is used in ISM, the pulse width is:

$$\begin{aligned} \operatorname{sech}^2\left[\frac{1}{l} \cdot x\right] &= \operatorname{sech}^2\left[\frac{1.76}{w_{FWHM}} \cdot x\right] \\ l &= \frac{w_{FWHM}}{1.76}. \end{aligned} \tag{6.7}$$

Amplitude is not affected by the temporal versus spatial measurements and therefore requires no transformation between the laboratory and NLTL lattice equations. When converting between the NLTL and KdV coordinates, however, the amplitude needs to be scaled by $4b$ [(2.8)].

When comparing the time evolution of a pulse in the canonical KdV system and the NLTL, the time also needs to be scaled by $24/v_0$ or $24\sqrt{LC_0}$.

The overall steps can by described as follows:

Laboratory → NLTL Lattice

$$\begin{array}{lcccc} w_t & & \rightarrow & & w_t \\ w_t & \rightarrow & w_t \cdot v_0\sqrt{\frac{3}{3-2bA}} & \rightarrow & w_n \\ A_t & & \rightarrow & & A_t \end{array} \tag{6.8}$$

NLTL → KdV Laboratory Approximation

$$\begin{array}{lcccc} w_t & & \rightarrow & & w_t \\ w_t & \rightarrow & w_t \cdot v_0\left(1 + \frac{bA}{3}\right) & \rightarrow & w_n \\ A_{t,n} & & \rightarrow & & A_{t,n} \end{array} \tag{6.9}$$

KdV Laboratory Approximation → KdV Approximation

$$\begin{array}{ccccc} n & & \rightarrow & & S = n - v_o t \\ t & & \rightarrow & & T \\ w_t & \rightarrow & w_t \cdot \frac{(1+\frac{bA}{3})}{\frac{bA}{3}} & \rightarrow & w_T \\ w_n & & \rightarrow & & w_S \\ A_{t,n} & & \rightarrow & & A_{T,S} \end{array} \tag{6.10}$$

KdV Approximation → Canonical KdV

$$\begin{array}{ccccc} w_S & \rightarrow & \frac{w_S}{1.76} & \rightarrow & l \\ A_{T,S} & \rightarrow & A_{T,S} \cdot 4b & \rightarrow & A_{\tau,x} \\ T & \rightarrow & T \cdot \frac{bA}{3\sqrt{LC_0}} & \rightarrow & \tau \end{array} \tag{6.11}$$

The overall transformation from laboratory coordinates to the KdV is summarized in Figure 6.3. Several aspects of each reference frame are worth noting. In the laboratory, the spatial data are discrete, not continuous, as in the other reference frames[1]. Another important observation is that the soliton is 3–5 sections wide, FWHM. This is useful for practical design as the approximate temporal width is on the order of $\sim 5 \cdot \sqrt{LC}$, where $\sqrt{LC}$ is the nominal time of a single section of the lattice[2].

There is very little difference between the NLTL lattice and the KdV laboratory approximation, as would be expected. When the KdV approximation is transformed into S and T, we note that the time axis changes. This is a result of the removal of the nominal velocity of the lattice, v_0. In essence the KdV approximation is a reference frame that moves at speed v_0. As a result the temporal width changes dramatically. This change in reference frames is noteworthy since many of the transient dynamics that we attribute to solitons, such as pulse sharpening and collisions, occur on this time frame. Several applications have proposed taking advantage of these transient dynamics for novel circuits and systems [17]. Unfortunately, since the transient dynamics, such as collision and pulse breakup, happen at a time scale much slower than the waveform propagation, $bAv_o/3$ vs. v_0 [(5.33)], a very long NLTL is needed. Damping, however, occurs in laboratory time and space, so that a long NLTL will result in large damping.

---

[1] One could plot n as a discrete variable in the NLTL lattice and KdV laboratory approximation, however, the functions are continuous, so it is customary to treat n as a continuous variable

[2] This is of course an approximation, since C changes with voltage, but it is a good rough guide to pulse width.

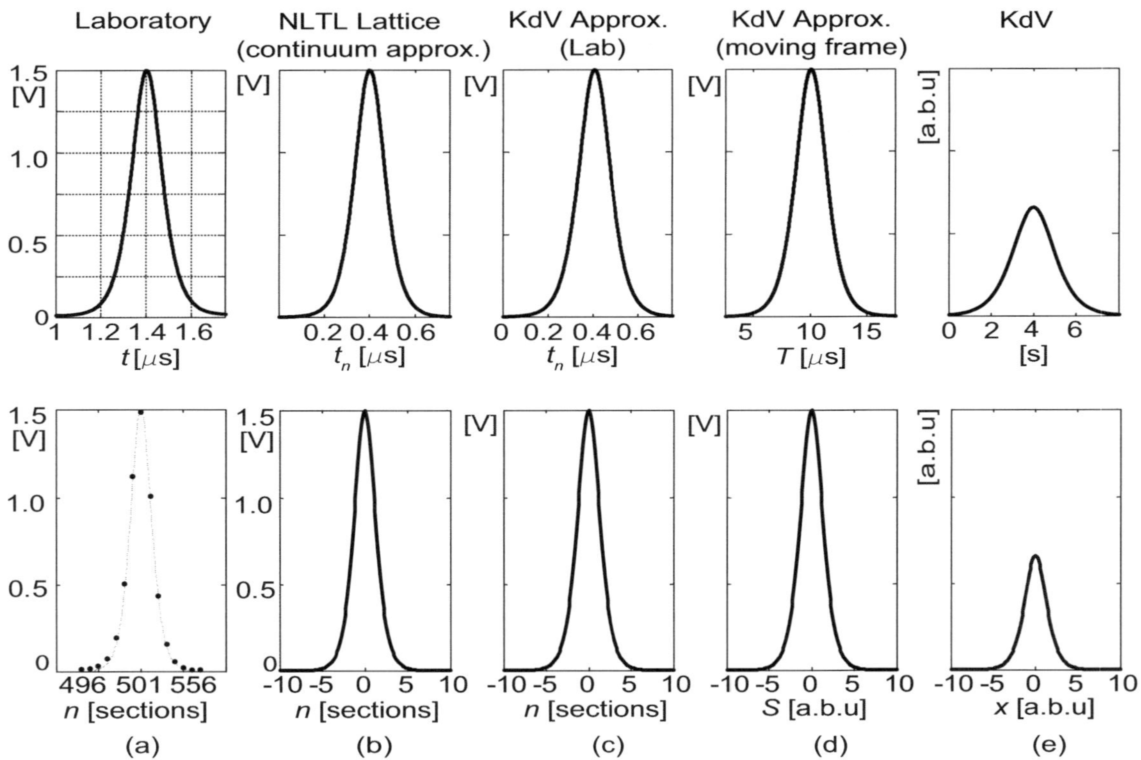

FIGURE 6.3
Transformation from laboratory coordinates to KdV coordinates. Data taken in a laboratory with an oscilloscope, (a), is converted into NLTL lattice coordinates, (b), using (6.2). Note that owing to the discrete sections of the NLTL, only discrete data points for the spatial waveform are possible (actual data shown by black circles). The faint grey curve is the NLTL lattice spatial waveform[(b) bottom], overlayed on the discrete lab data to show the correspondence. The NLTL lattice, (b), is then converted into the KdV laboratory approximation, (c). The KdV laboratory approximation, (c), is then converted from laboratory coordinates into S and T, (d). Finally, the KdV-approximation,(d), is converted

6.3 NLTL Characterization

In this section we will describe the characterization of the NLTL using two test beds. The first is an artificial NLTL, Fig. 5.1(b), constructed from discrete components. The second is a simulation model identical to the laboratory model, except for the inductor and resistors that were modeled as ideal (no parasitics) and the varactor as $C(V) = 140 \cdot (1 - 2 \cdot bV)$ pF. The simulation model allows us to decouple any parasitics from the fundamental dynamics.

We begin by determining the values of the discrete components. The inductor and capacitor were measured using an HP4275A *LCR* meter. Fig. 6.4 shows the capacitance of the varactor as a function of voltage. The extrapolated values for C_0 and b are 140 pF and 0.12, respectively. The capacitor's series resistance, R_C, is too small to be measured by the *LCR* meter; a *dc* measurement was made by plotting the forward bias current versus voltage of the pn junction varactor/diode using an Agilent 4156C Semiconductor Analyzer. The *dc* resistance measured was 1.48 Ω, which corresponds well with the capacitor Q of 80 at 50 MHz noted in the data sheet ($Q = \frac{1}{\omega RC}$)[3].

The inductors were also measured using the *LCR* meter. The inductance was found to be 23 μH over a frequency range of 1 to 10 MHz. The inductor's equivalent series resistance (*ESR*) at *dc* was approximately 0.65 Ω; over frequency, the *ESR* increased dramatically (note that *ESR* is different than R_L, in that *ESR* is an "equivalent" series resistance, where as R_L is the actual, physical series resistance). Figure 6.5 shows the measured *ESR* of the inductor versus frequency. This frequency dependence is likely due to the loss in the inductor's magnetic core as the operating frequency increases. For our simulation model, we arbitrarily used $R_L = 1.48$ ohms [see Fig. 5.3(a)], since our simple model of an inductor in series with a resistor does not allow for a frequency-dependent value.

Table 9.1 summarizes the components used and their key parameters: L, C_0, b, R_L and R_C.

In addition to determining the component parameters, the dispersion created by the periodicity of the NLTL can also be measured. Figure 6.6 shows the measured dispersion of the NLTL for small amplitude signals (50 mV). The wave number, k, was determined experimentally by measuring the wavelength in sections for one cycle. This is easily done by finding the frequency, whose wavelength is an integer number of sections.

[3]This Q would be taken at maximum reverse bias, or minimum capacitance.

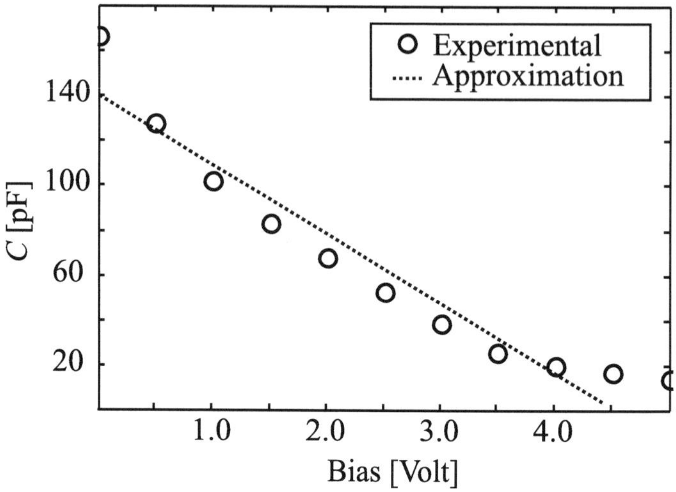

FIGURE 6.4
Capacitance versus bias voltage. Dashed line is linear approximation: $C(v) = C_0(1 - 2bV)$; $C_0 = 140$ pF and $b = .12$.

TABLE 6.1
Discrete NLTL components.

Component	Manufacturer	Part #	Parameter	Value
L	Panasonic	ELJEA220KF	L	23 μ H
L	Panasonic	ELJEA220KF	R_L	0.65 Ω (dc)
L	Panasonic	ELJEA220KF	ESR	freq. dependent
C	Zetex	ZMV934TA	C_0	140 pF
C	Zetex	ZMV934TA	b	0.11
C	Zetex	ZMV934TA	R_C	1.48 Ω
C	Zetex	ZMV934TA	G	0 Ω

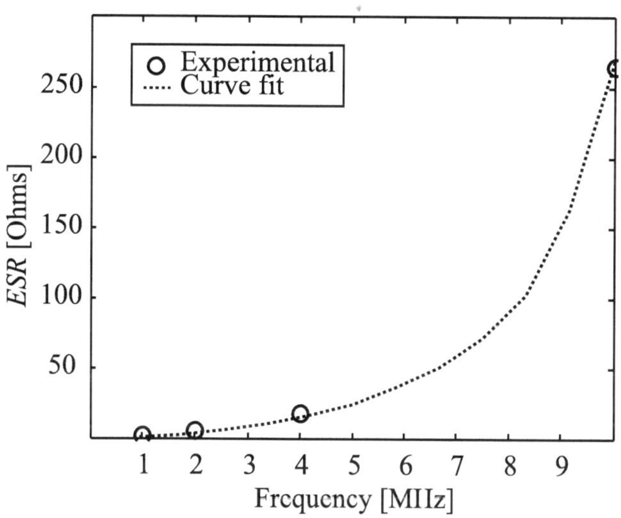

FIGURE 6.5
Inductor equivalent series resistance, ESR, versus frequency. Dashed line is a curve fit of data (LCR meter limited to 1, 2, 4 and 10 MHz measurement points)

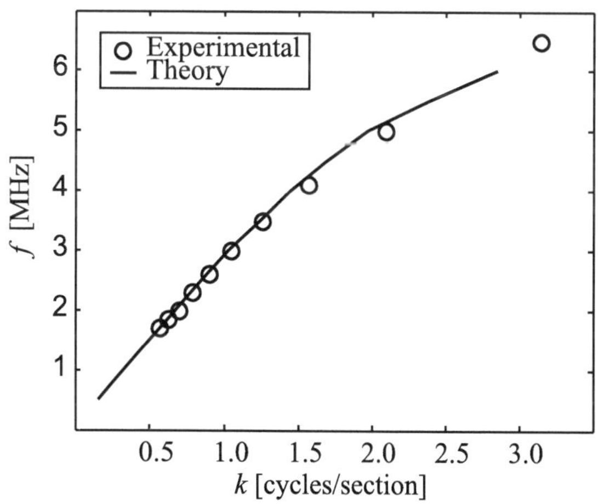

FIGURE 6.6
Measurement of the NLTL dispersion — f vs. k, where k is the wave number in 2π cycles per section.

The theoretical small-signal dispersion can be found from the linearized (5.27), where the term VV_S is negligible for small signals:

$$V_T + \frac{1}{24\sqrt{LC_0}} V_{SSS} = 0. \tag{6.12}$$

Assuming a propagating wave of the form, $e^{j(kS-\omega t)}$, and using $s = n - v_0 t$, where $v_0 = \frac{1}{\sqrt{LC_0}}$, the dispersion relationship in laboratory coordinates is:

$$\omega = kv_0 \left(1 - \frac{k^2}{24}\right). \tag{6.13}$$

The theoretical dispersion is plotted against the experimental data in Fig. 6.6 and shows an excellent match to the experimental data.

6.4 Inverse Scattering on the NLTL

In Sec. 3.1 in Chapter 2 we outlined the method for determining the emerging solitons from an initial profile. In this section we demonstrate this method by comparing experimental results with the predictions of the ISM. We use the NLTL described in the previous section for our experimental data. To test the largest combination of cases, the amplitude, A, was varied for a fixed pulse width, l, and then the pulse width was varied for a fixed amplitude.

An initial $A \cdot \text{sech}^2(x/l)$ profile was generated using and Agilent 33250 Arbitrary Waveform Generator. The waveform was generated in the time domain at the first node of the NLTL. The waveform was then converted from the NLTL reference frame to the KdV reference frame:

$$\begin{array}{ccccc} & & \textbf{NLTL} \rightarrow \textbf{KdV} & & \\ w_t & \rightarrow & w_t \cdot v_0 \left(1 + \frac{bA}{3}\right) & \rightarrow & w_n \\ w_n & \rightarrow & \frac{w_S}{1.76} & \rightarrow & l \\ A_{t,n} & \rightarrow & A_{t,n} \cdot 4b & \rightarrow & A_{\tau,x}. \end{array} \tag{6.14}$$

The eigenvalue of each initial profile was then calculated:

$$\kappa = \frac{1}{2l}((1 + 4Al^2)^{\frac{1}{2}} - (2n - 1)$$
$$\kappa > 0; \;\; n = 1, 2, 3, \ldots$$

From the soliton equation,

$$u(x,t) \;=\; 2\kappa^2 \cdot \mathrm{sech}^2(\kappa(x-4\kappa^2 t), (\kappa > 0) \qquad (6.15)$$

we can see that the amplitude is $2\kappa^2$ and that the spatial pulse width is $\frac{1.76}{\kappa}$. These are then scaled back to the NLTL domain:

KdV → NLTL

$$\begin{array}{lcccc} w_x = \frac{1.76}{\kappa} & \rightarrow & \frac{w_x}{v_0\left(1+\frac{bA}{3}\right)} & \rightarrow & w(t) \\ A_{\tau,x} = 2\kappa^2 & \rightarrow & \frac{A_{\tau,x}}{4b} & \rightarrow & A_{t,n} \end{array} \qquad (6.16)$$

This transformation introduces some small errors since the KdV soliton travels at a slightly different velocity than the NLTL soliton [See Fig. 5.2(b)].

Figures 6.7 and 6.8 show the results for the fixed amplitude and fixed width experiments, respectively. Note that the NLTL lattice width, w_{NLTL}, is a good fit while the KdV width, w_{KdV} is consistently larger. This is in agreement with Fig. 5.2(a), where we saw that the KdV width is slightly larger than the NLTL lattice for amplitudes around three volts.

In both figures there is a gray line marking the fixed amplitude or width of the input. The arrows show the regions where the pulse increases in amplitude and as a result narrows in width, and where the pulse decreases in amplitude and widens in width. These dynamics highlight the fact that the soliton has a unique form. If an input pulse is too narrow, it will grow into a fatter soliton. If it is too wide, it will sharpen into a narrower soliton. The widening or sharpening is such that the energy of the initial pulse is redistributed into a soliton of the proper form. During this formation process some energy is lost in a dispersive tail, i.e., not all of the energy is transferred to the emerging soliton.

6.5 Soliton Damping on the NLTL

In Sec. 5.5 we derived the damping of a soliton in a lossy medium and showed the change in amplitude over time for two specific loss mechanisms, inductor series resistance and capacitor series resistance. To demonstrate this in the lab system, we constructed in simulation an ideal NLTL with resistance in series with the inductor and capacitor

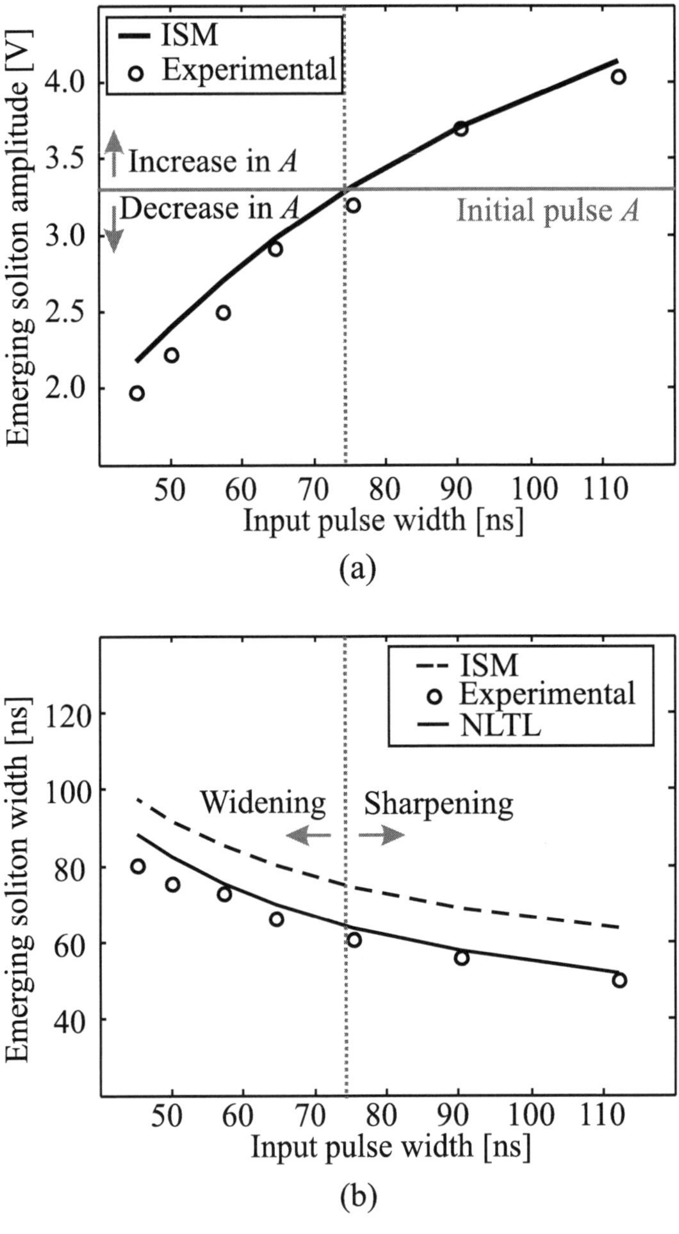

FIGURE 6.7
Experimental results of inverse scattering: amplitude fixed at 3.3 V, width varied from 45 ns to 110 ns. (a) Amplitude of emerging soliton. (b) Width of emerging soliton. Solid line is w_{NLTL}; dashed line is w_{KdV}.

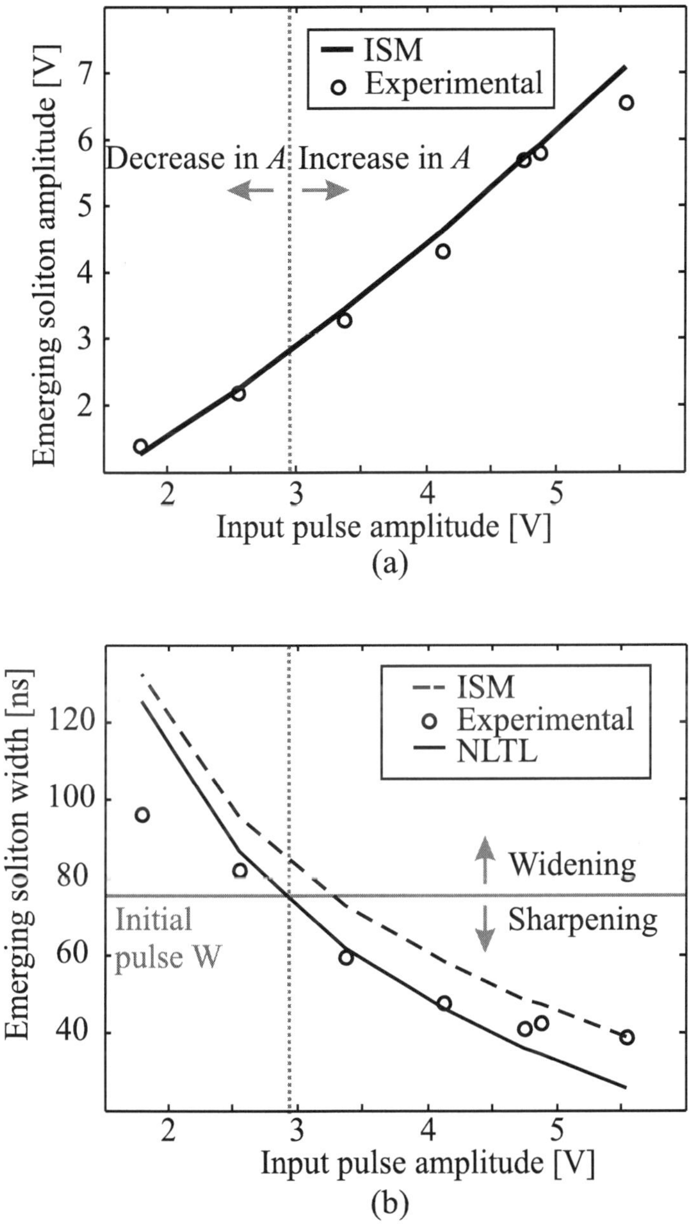

FIGURE 6.8
Experimental results of inverse scattering: pulse width fixed at 75 ns, amplitude varied from 1.8 V to 5.5 V (a) Amplitude of emerging soliton. (b) Width of emerging soliton. Solid line is w_{NLTL}; dashed line is w_{KdV}.

to demonstrate frequency-independent and dependent loss. The initial waveform was a 2.25 V soliton with a 116 ns pulse width, corresponding to the correct amplitude-width relationship of a soliton on the NLTL, (5.24).

Figure 6.9 shows the results of simulated damping of solitons on the NLTL. Plotted with each simulation result is a best-fit curve using the relationships (5.53) and (5.44). The excellent fit of the shape indicates that the mode of damping fits very well with theory. For each loss case, the value of ε was extracted from the best-fit curve. The values are a good fit to the theoretical values. The curves do differ at the beginning of the damping, which is due to an imperfect initial condition on the NLTL. As the pulse propagates and sheds its tail, it forms a soliton which damps according to theory. In Sec. 4.2 we showed that the the damping is an *exact* match for the KdV approximation. Here we show that it is a very good match for the NLTL system.

6.6 Numerical Accuracy

In the preceding section we matched the theoretical damping to the simulation results of an NLTL lattice. It should be noted that accurate simulation of the NLTL system requires high simulation tolerance, i.e., fine time step resolution. Figure 6.10 show simulation results for frequency-dependent loss with two tolerances. A significant difference in damping is seen, where the lower tolerance simulation gives an inaccurate result. Other inaccuracies have been noted when simulating the NLTL lattice with lower tolerances, in particular chaotic and unstable oscillations (See. Chapters 8 and 11) are *not* always observed. By increasing the tolerance, these dynamics become apparent.

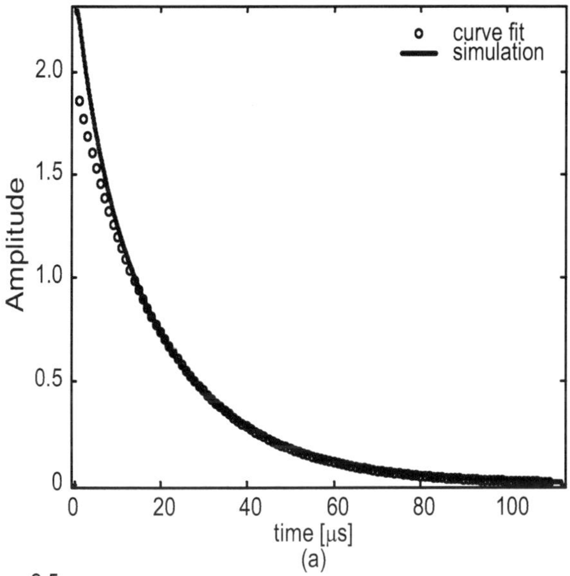

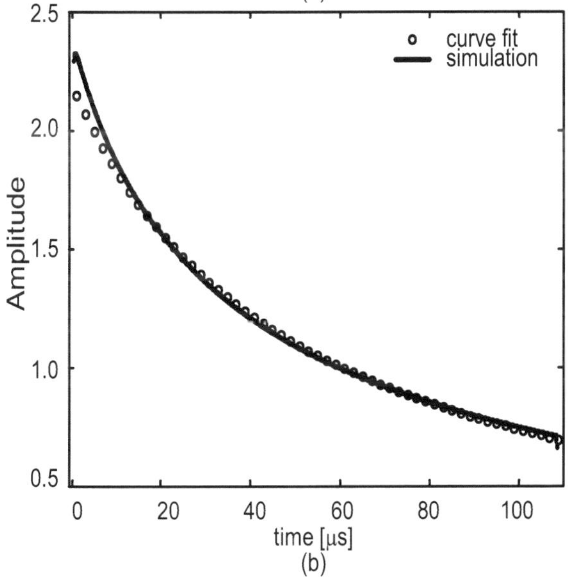

FIGURE 6.9
(a) Plot of soliton amplitude for frequency-independent loss. Curve fit using (5.44) shows that the soliton does damp with an exponentially decreasing amplitude. The calculated ε_2 from the curve fit is $3.6 \cdot 10^4$ which is close to the ideal $\frac{1}{2} \cdot (\frac{R_L}{L} + \frac{G}{C}) = 3.2 \cdot 10^4$. (b) Plot of soliton amplitude for frequency-dependent loss. Curve fit using (5.53) shows that the soliton does damp with a fractionally decreasing amplitude. The calculated ε_1 from the curve fit is $3.85 \cdot 10^4$ which is close to the ideal $\frac{R_C}{2L} = 3.2 \cdot 10^4$.

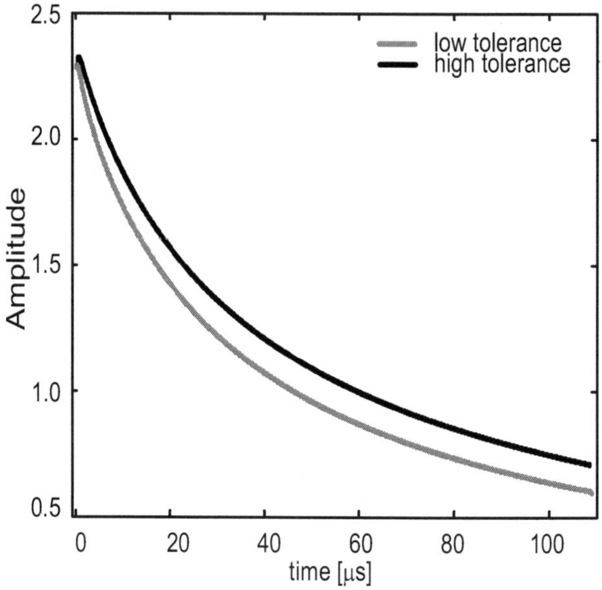

FIGURE 6.10
Plot of Fig. 6.9(b) for two tolerances: $reltol = 1 \cdot 10^{-3}$ (low) and $reltol = 1 \cdot 10^{-6}$ (high). The lower tolerance shows, incorrectly, a higher rate of damping.

Part III

Electrical Solitons: Application

7

NLTL as a Two-Port System

Xiaofeng Li
Harvard University

Michael W. Chen
Carnegie Mellon University

CONTENTS

In this chapter, we present applications of the NLTL as two-port (input and output) systems for pulse compression, shockwave formation, and harmonic generation. They have a wide range of applications including high-frequency sampling circuits, frequency synthesizer, high-precision time-domain reflectometry [13,14,56], and modulation pumping in mode-locked lasers.

7.1 Pulse Compression and Tapered NLTL

The transient soliton-forming process discussed in Chapters 2 and 3 can be utilized to compress an external high-frequency input signal into narrow soliton output pulses. Such pulse generation from electrical solitons on NLTLs was first proposed by Riley at Stanford in 1961. During the past 50 years, the active studies of pulse compression using NLTL have culminated in the state-of-the-art monolithic NLTL that can achieve a pulse width down to the low picosecond range (< 10 ps) [11,12,30,57]. The advantage of using the NLTL for narrow pulse generation is that the rise time of the pulses is no longer limited by the speed of transistors in active circuitry, but by the *RC* time constant of the varactors in

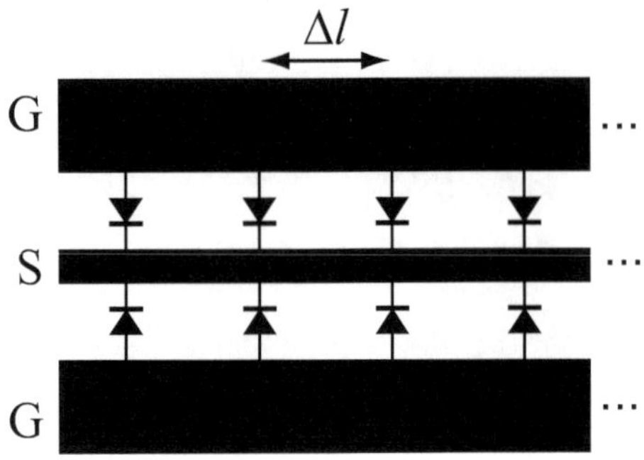

FIGURE 7.1
A monolithic NLTL implemented as a coplanar waveguide (CPW) periodically (spatial period = Δl) loaded with reverse-biased Schottky diodes.

NLTL, which are usually much faster than the transistors. In particular, Schottky diode varactors can be as fast as a couple of terahertz.

Figure 7.1 shows an NLTL implemented as a coplanar waveguide (CPW) periodically loaded with reverse-biased Schottky diodes. The CPW is fabricated monolithically on a semiconductor (e.g., GaAs and Si) substrate, in which the diodes/varactors are fabricated. Due to the periodic loading, the line exhibits a cutoff (Bragg) frequency, beyond which the wave cannot propagate. Since narrower pulses contain higher frequency contents, the cutoff frequency limits the minimum pulse width achievable using the NLTL. The cutoff frequency is given by

$$\omega_{\mathrm{Bragg}} = 2/\sqrt{LC}, \tag{7.1}$$

where L is the equivalent inductance of each waveguide section between two varactors, and C the average capacitance of the varactor (given the pulse amplitude) in parallel with the equivalent capacitance of the waveguide section. It is thus possible to achieve higher cutoff frequency, and hence sharper pulses, by using smaller varactors and finer sections.

From another perspective, solitons form and maintain their shape as a balance between nonlinearity and dispersion. For the large signal in pulse compression, the pulse typically experiences strong nonlinearity, so must substantial dispersion. This means that the soliton pulses naturally contain frequency contents towards the cutoff frequency, i.e., the high dispersion region as illustrated in Figure 7.2.

High pulse compression ratio, however, cannot be simply attained

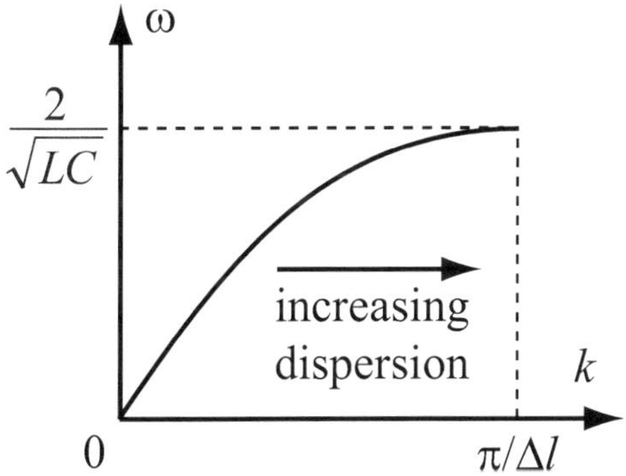

FIGURE 7.2
The dispersion curve of a periodically loaded line starts linear at low frequency and flattens as the wavelength and frequency approach the spatial period Δl and cutoff frequency $2/\sqrt{LC}$, respectively. The dispersion $|\partial^2\omega/\partial k^2|$ increases with the frequency.

by using an NLTL with high cutoff frequency. If a fat input pulse with width T_{pulse} enters an NLTL with cutoff frequency $f_c \gg 1/T_{\text{pulse}}$, it would inevitably break into multiple soliton pulses through the soliton transient process as discussed in Chapters 2 and 3. The typical output would be a primary pulse together with a series of secondary ones, which propagate at different speeds and would interfere with the primary pulse. For this reason, a maximum pulse compression ratio of only 2 can be achieved with a homogeneous (constant cutoff frequency) NLTL. Tan, Su, and Anklam (1988) at Hewlett-Packard introduced a tapered NLTL consisting of several cascaded homogeneous sections of NLTL with increasing cutoff frequency (i.e., decreasingly shorter line sections and smaller varactors) [32]. The basic idea can be seen by examining the soliton expression

$$V(n,t) = A \cdot \text{sech}^2\left[\sqrt{\frac{6bA}{3-2bA}}\left(n - v_0\sqrt{\frac{3}{3-2bA}}t\right)\right] \quad (7.2)$$

and noting that the temporal pulse width scales directly with $v_0 = 1/\sqrt{LC_0}$. By passing a pulse through an NLTL where v_0 is successively increased, the pulse width will be successively reduced. If the tapering and cutoff frequencies are chosen correctly, a single soliton will arise during the course of compression followed by a dispersive tail that represents the energy not converted into the single soliton. To illustrate this, we constructed in simulation several tapered NLTLs using the parameters of Chapter 6, Fig. 7.3. The NLTLs were tapered from $L = 23$ μH and $C = 140$ pF to $L = 4.5$ μH and $C = 30$ pF. Four tapering schemes were tried. The first was a simple abrupt transition from the initial LC values to the final LC values. The next three tapered the L and C, maintaining a constant impedance, $Z = \sqrt{L/C}$, using a linear, exponential and power series reduction. The results show a significant compression of the input into a narrow pulse at section 50. The exponential taper shows the best combination of sharp/tall output pulse and minimum tail. Tapering slowly, such as in the linear example, provides a clean pulse and minimum tail, however does not sharpen the pulse quickly. This is very important as the loss of the NLTL will begin to dominate and reduce the pulse amplitude and increase the pulse width. The power series taper and abrupt NLTL provide a narrow pulse quickly, however do not reduce the dispersive tail. The desired tapering method is dependant on the constraints of the actual NLTL. However, one can see that there is a natural tradeoff between quick, sharp pulse generation and minimization of the tail.

Pulse compression by a factor of 7 has been demonstrated with tapered NLTLs. Rodwell [57] and van der Weide [34] adapted Tan's technique and integrated it on a monolithic substrate with state-of-the-art GaAs technologies to produce the fastest electrical pulses ever generated from an all-electrical system. Rodwell's student, Case, achieved a pulse width of 5 ps [58] while van der Weide [34] demonstrated an edge rise time of 480 fs. These ultra-fast NLTLs have been used in numerous applications, in particular ultra-fast metrology, where a sampling bandwidth of over 700 GHz has been achieved [35].

Ideally, the NLTL can be tapered to arbitrary fineness with high cutoff (Bragg) frequency to achieve extremely sharp pulses. However, as the Bragg frequency increases, the pulse width will eventually be limited by the RC time constant formed by the series resistance and capacitance of the varactors (e.g., reverse-biased Schottky diodes), typically around a couple of THz. In fact, when the Bragg frequency exceeds the RC cutoff frequency, a shockwave instead of solitons will form through the transient wave-shaping process as the wave propagates along the NLTL. This will be the subject of next section.

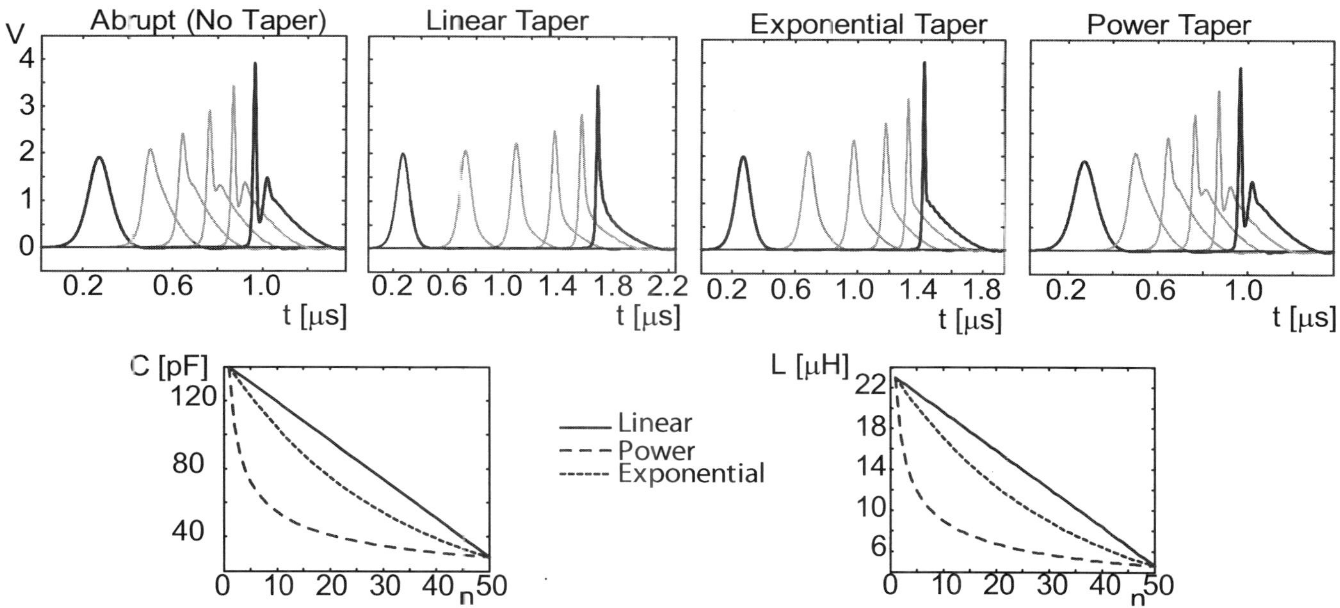

FIGURE 7.3
Simulation results of tapered NLTLs. The NLTL was constructed of 200 sections, with the first 50 sections tapered. Initial and final waveforms (section 50) are shown in black while intermediate waveforms are shown in gray and represent the waveforms at 10, 20, 30 and 40 sections. Four tapering schemes were implemented: abrupt, linear, exponential and power series. The capacitance and inductance values for each section are also shown.

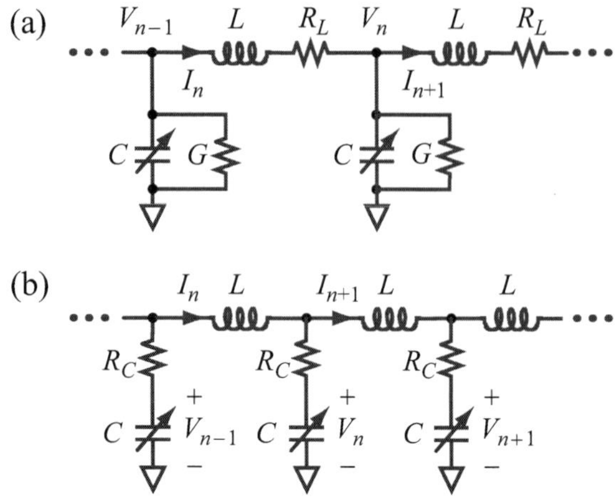

FIGURE 7.4
Loss mechanisms in NLTL: (a) Frequency-independent loss due to serial loss of the inductors and parallel loss of the capacitors; (b) Frequency-dependent loss due to the series resistance of the varactors.

7.2 Shockwave NLTL

In Chapters 2 and 3, we briefly discussed two loss mechanisms in NLTL. The first one arises from the serial loss of the inductors and parallel loss of the capacitor, and is independent of frequency [Fig. 7.4(a)]. The other arises from the series resistance of the varactors (e.g., contact resistance of the Schottky diodes), and increases with the frequency [Fig. 7.4(b)]. The latter introduces another cutoff frequency to the NLTL, which is determined by the RC time constant of the varactors. Unlike the Bragg cutoff frequency, ω_{Bragg}, which arises from dispersion, this RC cutoff frequency, ω_{RC}, is dissipative in nature, meaning that frequency components beyond ω_{RC} are heavily attenuated as they propagate along the line. When $\omega_{\text{Bragg}} \ll \omega_{RC}$, dispersion dominates over dissipation, and soliton pulses are formed through the transient soliton process. When $\omega_{RC} \ll \omega_{\text{Bragg}}$ and dissipation dominates, solitons cannot form as their high frequency contents toward ω_{Bragg} are heavily attenuated due to the frequency-dependent loss. Instead, a shockwave will form, which features a steep rising or falling edge, and can be useful in high-speed sampling circuits, e.g., to strobe diode sampling bridges used in microwave network analyzers and oscilloscopes. Subpicosecond falling edge [34] has

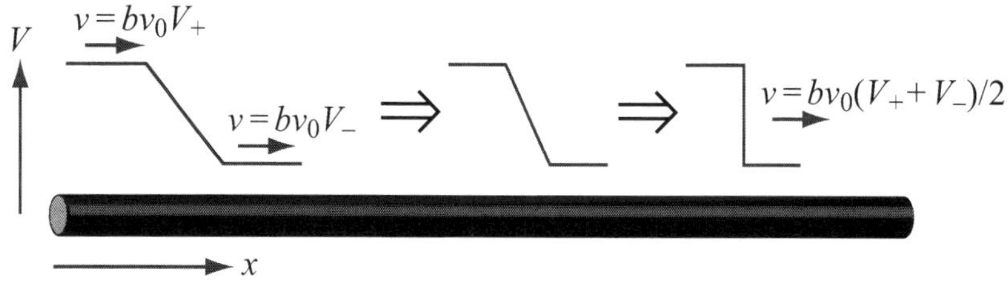

FIGURE 7.5
Shockwave formation of uni-directional nonlinear waves.

been achieved with monolithic GaAs NLTLs, which potentially extends the sampling circuit and instrument bandwidth to beyond 1 THz.

Before going into details of shockwave on NLTL, let's first have a quick look at the uni-directional nonlinear wave to get the basic idea of shockwaves. The uni-directional nonlinear wave is described by

$$V_t + bv_0 V V_x = 0, \tag{7.3}$$

where the wave speed $v = bv_0 V$ depends on the amplitude V. Consider a step-like wave. The high amplitude part propagates faster than the low amplitude part. As the high amplitude part catches up with the low amplitude part, the falling edge of the step becomes steeper and steeper, and eventually a shockwave forms (Fig. 7.5). The shockwave propagates at the average speed of the higher and lower levels, or $v = bv_0(V_+ + V_-)/2$.

Now, let's have a closer look at the shockwave in NLTL. We repeat the first part of the analysis in Sec. 5.5 for continuity and to develop the derivation of the shock wave solution for large dissipation. Using the circuit diagram in Fig. 7.4(b), we write down the nodal equations,

$$\begin{aligned} I_n - I_{n+1} &= \frac{\partial Q(V_n)}{\partial t}, \\ L\frac{\partial I_n}{\partial t} &= \left[V_{n-1} + R\frac{\partial Q(V_{n-1})}{\partial t}\right] - \left[V_n + R\frac{\partial Q(V_n)}{\partial t}\right], \\ L\frac{\partial I_{n+1}}{\partial t} &= \left[V_n + R\frac{\partial Q(V_n)}{\partial t}\right] - \left[V_{n+1} + R\frac{\partial Q(V_{n+1})}{\partial t}\right]. \end{aligned} \tag{7.4}$$

Note that, for convenience, we have chosen V_n as the voltage on the nonlinear capacitor. Combining these, using the charge-voltage relation of the varactor and expanding the voltage in a Taylor series, we obtain (See Sec. 5.5):

$$V_{tt} - b(V^2)_{tt} = v_0^2 \left(V_{nn} + \frac{1}{12} V_{nnnn} \right) + \frac{R}{L} V_{nnt}, \tag{7.5}$$

where $v_0 = 1/\sqrt{LC_0}$. Note that, in the last term, we have kept only the dominant effect of dissipation. Finally, we utilize the reductive perturbation method [54], with $V \to \varepsilon V$, $R/L \to \varepsilon^{1/2}(R/L)$, and a scaled moving frame S and time T

$$\begin{cases} S = \varepsilon^{1/2}(n - v_0 t) \\ T = \varepsilon^{3/2} t \end{cases} \Rightarrow \begin{cases} \dfrac{\partial}{\partial n} = \varepsilon^{1/2} \dfrac{\partial}{\partial S} \\ \dfrac{\partial}{\partial t} = -\varepsilon^{1/2} v_0 \dfrac{\partial}{\partial S} + \varepsilon^{3/2} \dfrac{\partial}{\partial T} \end{cases} \tag{7.6}$$

and keeping terms up to $O(\varepsilon^3)$, to reduce (7.5) into a simpler form

$$V_T + b v_0 V V_S + \frac{v_0}{24} V_{SSS} - \frac{R}{2L} V_{SS} = 0. \tag{7.7}$$

Equation. (7.7) is known as the KdV-Burger's equation, which can be considered as the uni-directional nonlinear wave (7.3) with additional dispersion (V_{SSS}) and frequency-dependent dissipation (V_{SS}) terms. In absence of the loss term V_{SS}, (7.7) reduces to the KdV equation that we have studied in Chapter 5. Indeed, when $\omega_{\text{Bragg}} \ll \omega_{RC}$, the dispersion term V_{SSS} dominates, and soliton pulses form and get slowly damped by the dissipation term V_{SS} as discussed in Chapter 5. On the other hand, when $\omega_{RC} \ll \omega_{\text{Bragg}}$, the dissipation dominates over dispersion as $v_0 \ll R/L$. The equation thus becomes the Burger's equation, if the V_{SSS} term is ignored.

Similar to what we have done in Chapter 2 and 5, we seek a traveling wave solution to (7.7) of the form $V = V(S - cT)$, propagating at velocity c (in the moving frame). With $\xi = S - cT$, $\partial/\partial S = d/d\xi$ and $\partial/\partial T = -c d/d\xi$, (7.7) becomes an ordinary differential equation (ODE):

$$-cV' + b v_0 V V' + \frac{v_0}{24} V''' - \frac{R}{2L} V'' = 0, \tag{7.8}$$

where the prime represents $d/d\xi$. Integrating once with the boundary condition $V \to V_\pm$ as $\xi \to \mp\infty$, we have

$$-cV + \frac{b v_0}{2} V^2 + \frac{v_0}{24} V'' - \frac{R}{2L} V' = A, \tag{7.9}$$

with integration constant $A = -b v_0 V_+ V_- / 2$, and propagation speed

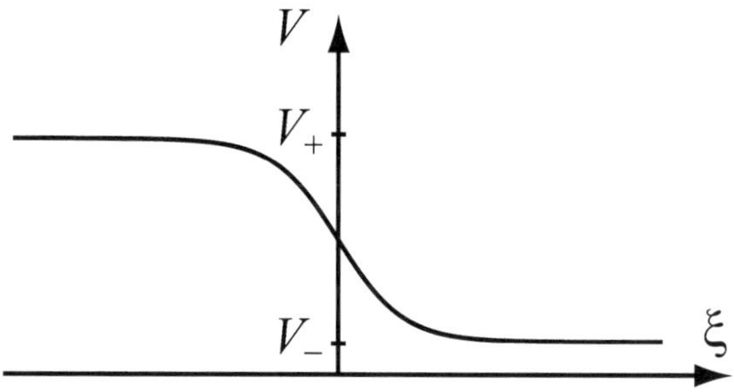

FIGURE 7.6
Shockwave solution of the Burger's equation, where the dispersion term V_{SSS} has been ignored. It propagates at the same speed as the shockwave of uni-directional nonlinear wave, but has a nonzero rise time.

$c = bv_0(V_+ + V_-)/2$. Note that this propagation speed is the same as that of the shockwave in uni-directional nonlinear waves (Fig. 7.5).

Equation (7.9) is a 2nd order nonlinear ODE, which is not straightforward to solve, but can be understood qualitatively. To begin with, if the 2nd-order term V'' is ignored when $\omega_{RC} \ll \omega_{\text{Bragg}}$, the resulting 1st-order equation can be solved exactly, yielding (Fig. 7.6)

$$V = \frac{V_+ + V_-}{2} - \frac{V_+ - V_-}{2} \tanh\left[\frac{bv_0 L}{2R}(V_+ - V_-)\xi\right], \tag{7.10}$$

or, in terms of the original laboratory coordinates:

$$V = \frac{V_+ + V_-}{2} - \frac{V_+ - V_-}{2} \tanh\left[\frac{bv_0 L}{2R}(V_+ - V_-)(n - (v_0 + c)t)\right]. \tag{7.11}$$

In time domain, this gives a step function with a rising edge (10 % to 90 %) of

$$t_{\text{rise}} \approx \frac{4.4RC}{b(V_+ - V_-)}, \tag{7.12}$$

which is proportional to the RC time constant, and decreases with increasing nonlinearity b and step size $(V_+ - V_-)$.

This step-like shockwave propagates without change of shape, and is sometimes known as a kink soliton. It results from the balance of nonlinearity and dissipation (c.f. dispersion), in that the high frequency components of the kink are constantly dissipated due to the frequency-dependent loss, while the energy is continuously converted from lower to higher frequencies by the nonlinearity. Note that the energy loss is compensated by the infinite energy supply from the step-like waveform.

To understand the effect of the dispersion term (V_{SSS} or V''), we rewrite (7.9) into two coupled 1st-order ODEs by introducing $W \equiv V'$:

$$\begin{aligned} V' &= W, \\ W' &= \frac{24}{v_0}\left(A + cV - \frac{bv_0}{2}V^2 + \frac{R}{2L}W\right). \end{aligned} \tag{7.13}$$

It can be easily shown that (7.13) has two equilibrium points at $(V, W) = (V_{\pm}, 0)$. In the phase space formed by V and W, the solution to (7.13) would be a trajectory starting from $(V_+, 0)$ and ending at $(V_-, 0)$; and the solution's behavior can be determined by the trajectory's characteristic around the two points (Fig. 7.7). We thus linearize (7.13) about the equilibrium points to obtain

$$\begin{bmatrix} \Delta V \\ \Delta W \end{bmatrix}' = \begin{bmatrix} 0 & 1 \\ \mp 12b(V_+ - V_-) & 12RCv_0 \end{bmatrix} \begin{bmatrix} \Delta V \\ \Delta W \end{bmatrix}, \tag{7.14}$$

where the two eigenvalues of the 2×2 matrix are

$$\lambda = 6RCv_0 \pm \sqrt{(6RCv_0)^2 - 12b(V_+ - V_-)} \qquad \text{for } (V, W) = (V_+, 0),$$

$$\lambda = 6RCv_0 \pm \sqrt{(6RCv_0)^2 + 12b(V_+ - V_-)} \qquad \text{for } (V, W) = (V_-, 0).$$

The eigenvalues for $(V, W) = (V_-, 0)$ are always real. However, the eigenvalues for $(V, W) = (V_+, 0)$ are real only when $RCv_0 \geq \sqrt{b(V_+ - V_-)/3}$ or $\omega_{\text{Bragg}} \geq \omega_{RC}\sqrt{4b(V_+ - V_-)/3}$. If ω_{Bragg} is not large enough, the dispersion will generate imaginary eigenvalues, meaning that the shockwave will have oscillating ripples about V_+ as the signal steps up from V_- to V_+.

We illustrate shockwave formation on an NLTL by using our prototypical NLTL with $b = 0.11$, $L = 23$ μH and $C = 140$ pF. The input is a 2 V, 2.4 μs rise time, step function. We simulate three conditions of increasing ω_{RC}, Fig. 7.8, where R_C is the resistor in series with the nonlinear capacitor. As ω_{RC} approaches ω_{Bragg}, the ripples are reduced; however, so is the sharpening of the edge. The condition, $\omega_{\text{Bragg}} \geq \omega_{RC}\sqrt{4b(V_+ - V_-)/3} = 0.54{\cdot}\omega_{RC}$ or $\omega_{RC} \sim 2{\cdot}\omega_{\text{Bragg}}$, is shown on the right-hand side of Fig. 7.8, where one can see that indeed the oscillations, or imaginary eigenvalues, have been eliminated. It is the designer's choice to create the right ω_{RC}-ω_{Bragg} relation to obtain the desired edge sharpening and shockwave.

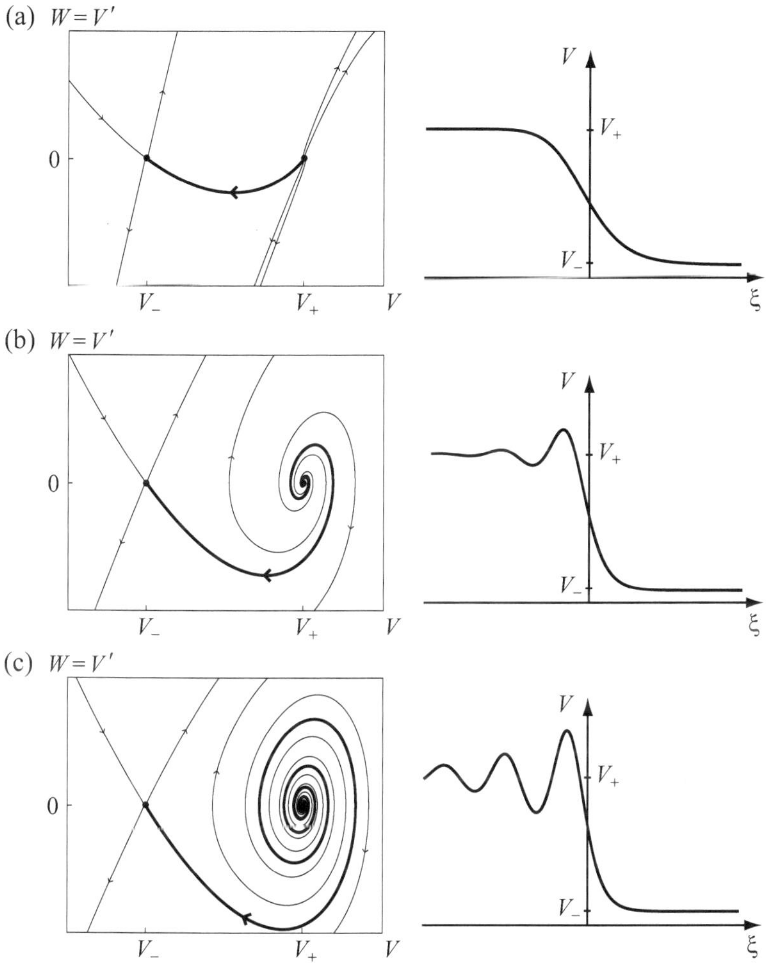

FIGURE 7.7
Shockwave solution of the KdV-Burger's equation (7.9) with increasing dispersion (decreasing ω_{Bragg}). Left column: phase space diagrams, in which the bold trajectories from $(V_+, 0)$ to $(V_-, 0)$ correspond to the solutions; right column: shockwave solutions to (7.9). Note that, in time domain, the waveform will be a rising step from V_- to V_+.

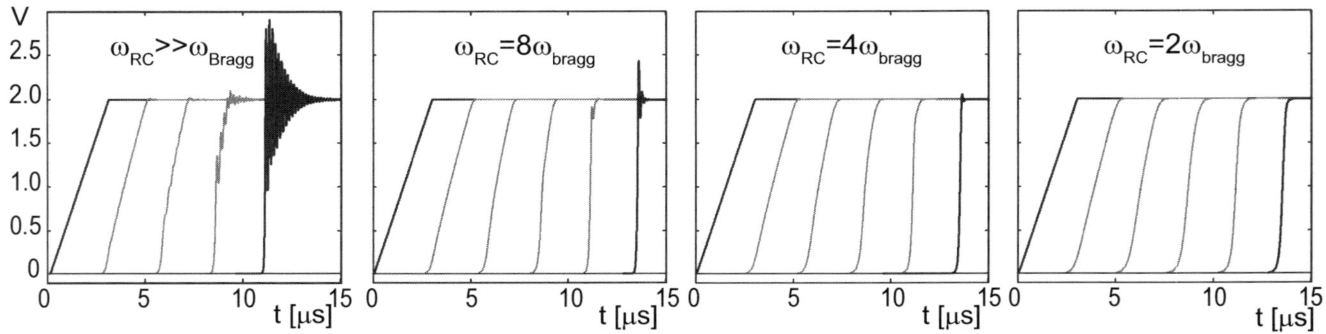

FIGURE 7.8
Simulation results of shockwave formation on the lossy NLTL. Four NLTLs were simulated with R =0, 50, 100 and 200 Ω. Each plot shows the waveform a sections 0, 10, 20, 30, 40 and 50, where the initial and final waveforms are in black. The rise times were 51 ns, 97 ns, 149 ns and 282 ns, representing an edge sharpening of 47×, 25×, 16× and 9×, respectively.

7.3 Harmonic Generation

In the previous two sections, we have discussed two applications of NLTLs for pulse compression and step function generation with extremely short rise and/or fall time. The resulting sharp pulses and steep step functions contain high frequency harmonics beyond the bandwidth of the active transistors, thanks to the nonlinearity of NLTLs. It is thus natural to utilize this nonlinearity for harmonic generation, in which the NLTL takes a sinusoidal signal (instead of pulses or step functions) as input and produces harmonics at higher frequencies.

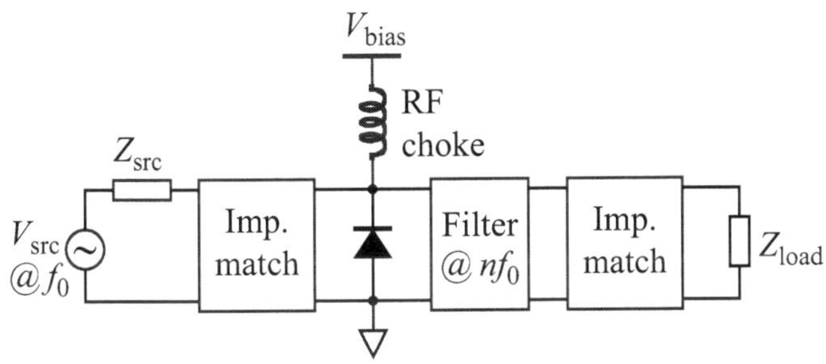

FIGURE 7.9
Harmonic generation using discrete nonlinear devices (e.g., Gunn diodes).

In fact, any nonlinear device can be utilized for harmonic generation. Fig. 7.9 shows a scheme widely used in practice, such as the Gunn diode frequency multipliers, commonly used for millimeter/sub-millimeter wave generation. For maximum power conversion, impedance matching networks are generally used to match the source (Z_{src}) and load (Z_{load}) to the capacitive impedance of the diode. The narrow bandwidth of these impedance matching networks, however, limits the frequency range over which efficient harmonic generation can be obtained. By distributing the nonlinear capacitors in an NLTL, efficient harmonic generation can be achieved over a wide bandwidth up to the Bragg cutoff frequency, and the line impedance given by $\sqrt{L/C}$ can be matched to the source and load by a proper design of the line geometry and varactor size.

In harmonic generation, the NLTL is optimized for producing the desired higher order (e.g., 2nd, 3rd or 5th) harmonics. The soliton-breaking process, which is undesirable in pulse compression, may be promoted for high-efficiency conversion to certain harmonics. For example, second order harmonics can be efficiently generated, when $\omega_{\mathrm{Bragg}}/3 < \omega_{\mathrm{in}} < \omega_{\mathrm{Bragg}}/2$. As shown in Fig. 7.10, each peak of the input sinusoidal wave breaks into two solitons, and the output waveform contains a strong 2nd-harmonic component. This can also be understood in the frequency domain: higher order harmonics cannot be generated because their frequencies are beyond the Bragg cutoff frequency of the NLTL; consequently, energy conversion happens primarily between the fundamental tone and the 2nd harmonic. Specialized NLTL structures have also been fabricated for highly efficient harmonic generations [30, 59].

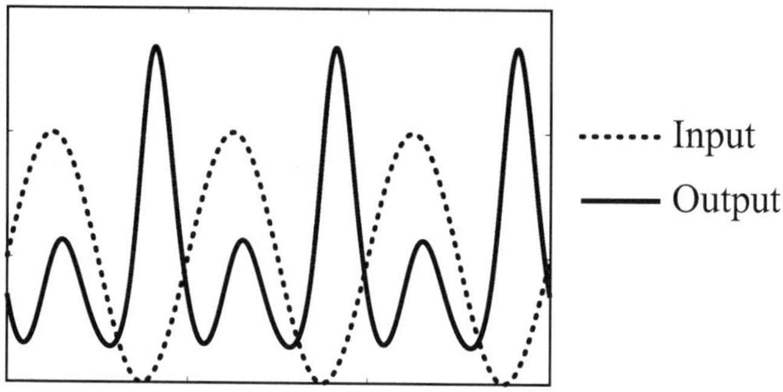

FIGURE 7.10
Second order harmonic generation using NLTL with $\omega_{\mathrm{in}} \approx \omega_{\mathrm{Bragg}}/2.5$. A conversion efficiency of 64 % (−1.9 dB) has been achieved. Note that each peak of the sinusoidal input breaks into two soliton pulses.

7.4 Summary

In this chapter, we describe three applications of electrical solitons in NLTLs as two-port networks. The central idea is to utilize the nonlinearity and broad bandwidth of the NLTL to generate sharp pulses, steep step functions, and harmonic contents, with frequencies beyond the bandwidth of active transistors. However, the disadvantage of this two-port topology is that it requires an external high-frequency input, whose quality and waveform can greatly affect the output. It is therefore necessary to introduce feedback in order to achieve robust operation. Along this line, self-sustained electrical soliton oscillators have been developed, which will be the topic of the following sections.

8

The Soliton Oscillator

CONTENTS

This chapter introduces the concept of the electrical soliton oscillator. It begins with a brief description of the general topology and then discusses the difficulties with constructing a stable electrical soliton oscillator using standard techniques. The remaining portion of this chapter describes the solution for overcoming these challenges. Verification of these methods and the general topology will be presented in the succeeding chapters.

8.1 Basic Topology

We begin by considering a closed-loop (ring) NLTL as shown in Fig. 8.1(a). If only unidirectional propagation is allowed, the possible soliton propagation modes (cnoidal wave modes) on the ring are the ones that satisfy the periodic boundary condition $l = n \cdot L$ [$n = 1, 2, 3, \ldots$, i.e., (2.27)] where l and L are the circumference of the ring and the spacing between neighboring solitons, respectively [For L, refer to (2.29)]. The

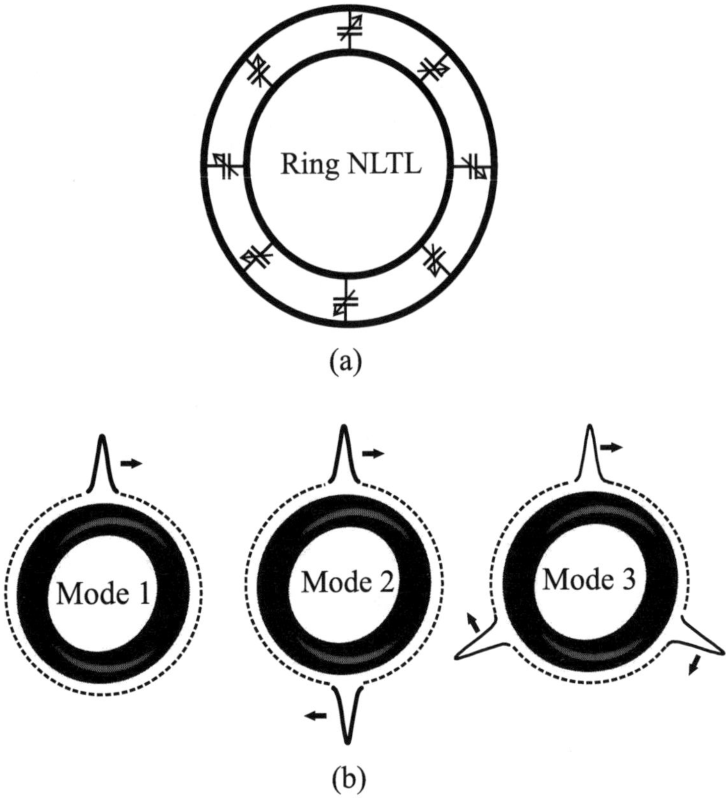

FIGURE 8.1
(a) Ring NLTL structure. (b) Modes of the ring NLTL

first three soliton propagation modes formed on the ring are shown in Fig. 8.1(b).

Our starting idea to construct an electrical soliton oscillator was to break the ring NLTL and insert a non-inverting amplifier as shown in Fig. 8.2. The purpose of the amplifier is to enable initial startup from noise and to compensate for system loss in steady state as is commonly done in *LC* or standing wave oscillators, e.g., [60,61]. The ultimate goal is for the circuit of Fig. 8.2 to self-generate and self-sustain one of the soliton modes of the ring NLTL.

As will be seen shortly, the proposed topology will indeed lead to an oscillation. However, the oscillation stability depends strongly on the amplifier characteristics. For the oscillator to generate a stable soliton pulse train, it is essential for the amplifier to "tame" the complex soliton

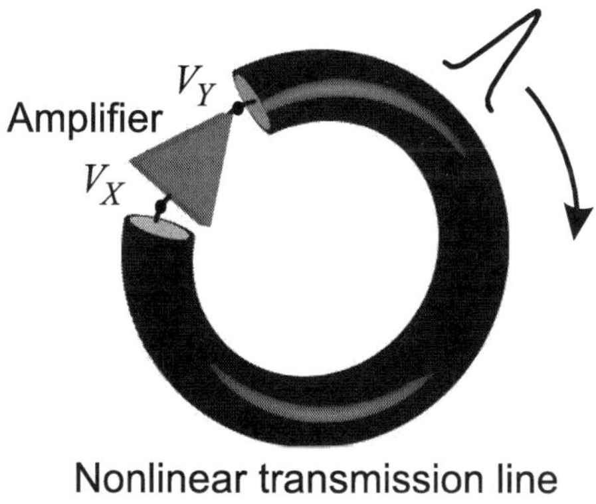

FIGURE 8.2
Soliton oscillator topology.

dynamics of the NLTL, which, unfortunately, is not the case for standard amplification techniques.

8.2 Instability Mechanisms

In this section, we examine the dynamics of Fig. 8.2 with two commonly-used amplifiers. This will lead us to identify the instability mechanisms associated with the oscillator, which is key to constructing a stable soliton oscillator.

8.2.1 Case I — Voltage Limiting Amplifier

Let us first consider the case where a standard voltage limiting, non-inverting amplifier is used in the circuit of Fig. 8.2. The transfer function of this amplifier is shown in Fig. 8.3(a). The amplifier is biased at a fixed operating point. Behavioral simulations using MATLAB show that the circuit of Fig. 8.2 indeed self-starts into oscillation, but the oscillation is unstable with significant amplitude and pulse repetition variations, often tending toward what appears to be a chaotic state, as shown in Fig. 8.3(b).

This oscillation instability arises from the signal clipping in the amplifier in connection with the unique NLTL properties. To see this, assume that a soliton pulse appears at the input of the amplifier at a certain time [V_X on the left of Fig. 8.3(c)]. This soliton pulse, after passing through the amplifier, will turn into a square pulse due to the clipping of the amplifier [V_Y in Fig. 8.3(c)]. As explained in Fig. 3.6, this square pulse will break up into several solitons with different amplitudes while propagating down the NLTL [right-most waveform in Fig. 8.3(c)]. The multiple soliton pulses will travel at different speeds, eventually appearing again at the input of the amplifier. This process repeats itself, creating many soliton pulses with various amplitudes in the loop. These soliton pulses propagate with different speeds and continually collide with one another, causing time (phase) shifts and amplitude variations, making the oscillation unstable.

8.2.2 Case II — Linear Amplifier

The discussion above shows that signal distortion has a negative impact on the oscillation stability, suggesting that one *might* be able to attain a stable soliton oscillation if signal distortion is mitigated. Ballantyne, et al. [36, 62, 63] implemented such a system using a linear amplifier whose transfer function is shown in Fig. 8.4(a). By using the linear amplifier and adding additional frequency-dependent loss to the NLTL, he could produce a periodic soliton pulse train as illustrated in Fig. 8.4(b). In the same system, however, other oscillations uncontrollably appeared with slight changes in gain, termination, or with external perturbations, indicating a lack of robustness. These waveforms sometimes contained multiple solitons with varying amplitudes/spacings and pulses that continuously moved relative to one another and collided (see Fig. 8.4(c) for a depiction). While Ballantyne's system is invaluable for providing an opportunity to examine soliton dynamics, it is not able to reproduce a stable soliton pulse train, i.e., a robust oscillation.

8.3 Identification of Three Instability Mechanisms

Ballantyne's work suggests that distortion reduction is a necessary, but not a sufficient, condition to completely stabilize the oscillation and ensure reproducibility. There are two other important instability mechanisms that we overlooked in the previous arguments.

First, perturbations arising from inherent noise and from inhomogeneities on the NLTL (e.g., amplifier, termination) can excite various solitons in addition to the existing, desired soliton pulse train. The non-

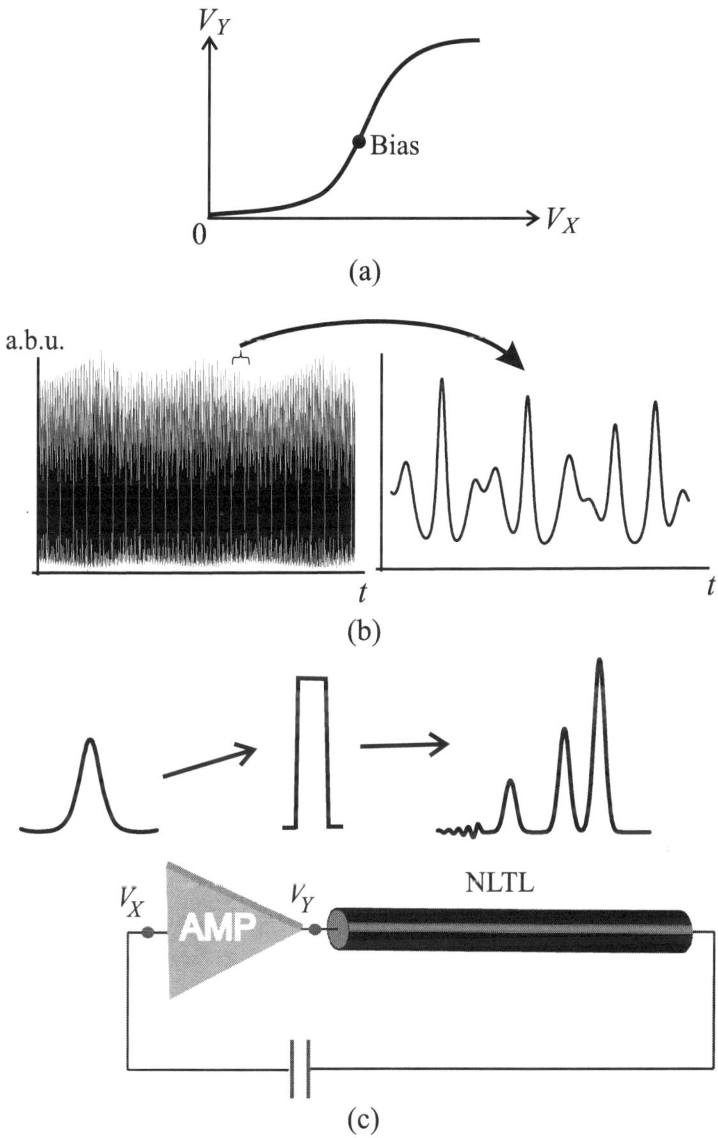

FIGURE 8.3
(a) Nonlinear transfer function of a voltage limiting amplifier. (b) MATLAB behavioral simulation results of an unstable oscillation of Fig. 8.2 with the voltage limiting amplifier. (c) Impact of signal clipping, or distortion.

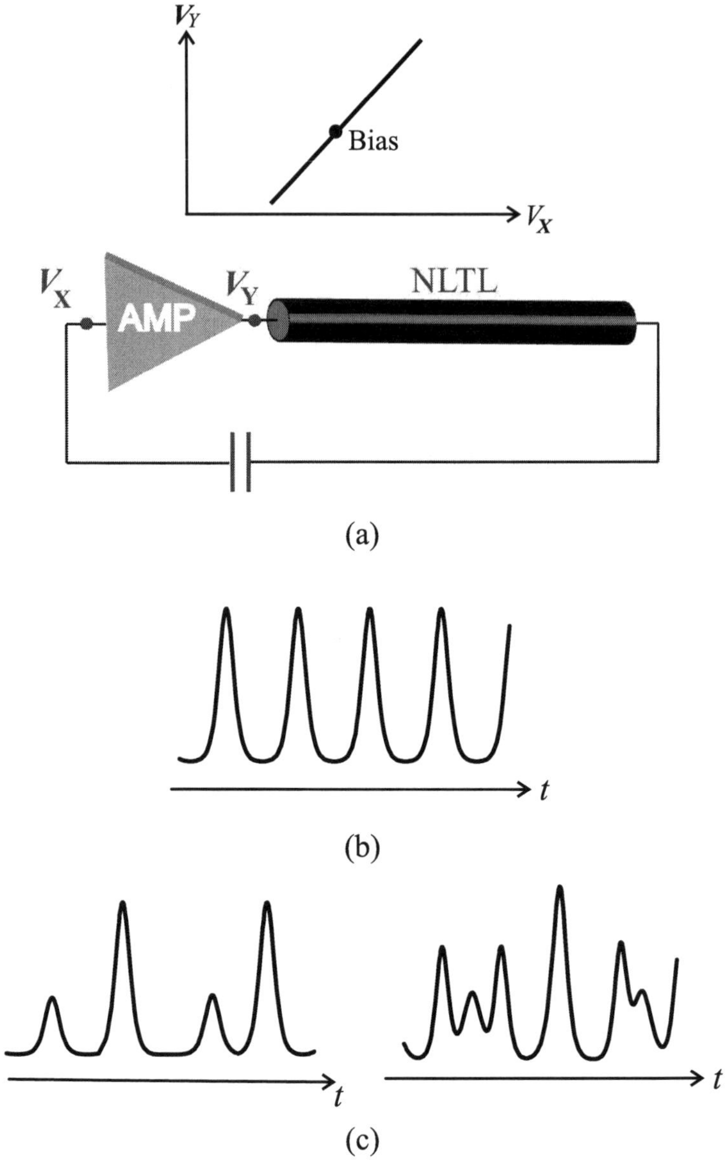

FIGURE 8.4
(a) Linear amplifier transfer function. (b) Soliton oscillation of Fig. 8.2 with the linear amplifier [36, 62, 63]. (c) Depiction of other possible output waveforms [63].

linear collisions between the perturbing solitons and the existing soliton pulse train result in significant time (or phase) shifts and amplitude variations, leading to unstable oscillations. Second, unless it is ensured that a single mode is generated every time (see Fig. 8.1 for possible mode examples), nonlinear inter-mode collisions on the NLTL could lead to instability.

Summarizing, to attain a stable, robust soliton oscillation with the circuit of Fig. 8.2, the amplifier should possess at least the following three capabilities:

1. Distortion reduction
2. Perturbation rejection
3. Single mode selection.

Clearly the voltage limiting amplifier does not satisfy Condition (1). Neither the voltage limiting amplifier nor the linear amplifier satisfies (2) or (3), which will become evident in the following section.

8.4 NLTL Soliton Oscillator — Working Model

Our approach to attain a stable, robust soliton oscillation with the circuit of Fig. 8.2 was the development of a unique amplifier. The amplifier incorporates an adaptive bias control that exploits the amplifier's nonlinear transfer function. This approach allows the simultaneous satisfaction of the three stability requirements mentioned above. This section discusses the operating principles of our amplifier/oscillator.

8.4.1 Operating Principles

Our self-sustained soliton oscillator is schematically illustrated in Fig. 8.5. It is exactly the same topology as originally proposed in Fig. 8.2 and interestingly uses the voltage limiting amplifier that was shown with Fig. 8.3 to cause apparently chaotic oscillations. The critical difference in our amplifier is that instead of allowing the nonlinearity to destabilize the oscillation, we take advantage of it to provide stability. This is done by using an adaptive bias control to move the amplifier bias point as the *dc* component of the amplifier output, V_Y in Fig. 8.5, changes.

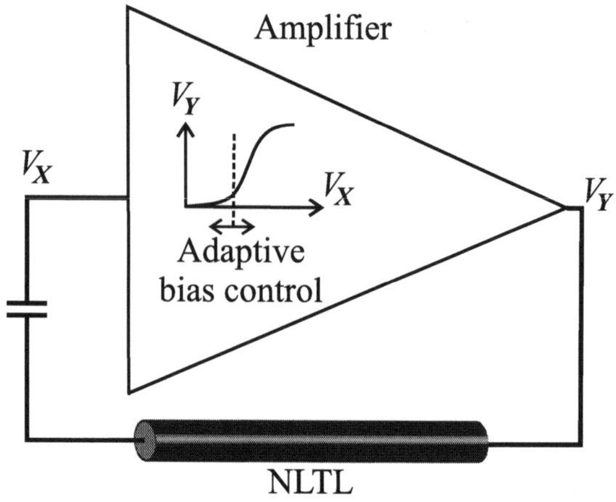

FIGURE 8.5
Our soliton oscillator with the nonlinear amplifier that performs the adaptive bias control, providing the key stabilizing mechanisms.

8.4.1.1 Bias adjustment

The bias of the amplifier is adjusted by using the change in *dc* component of the output, V_Y, as the oscillation grows from a low amplitude signal at start-up to a large signal, pulse-based oscillation in steady state, Fig. 8.6. Other methods may be used to make this bias adjustment, such as output peak-detection, however, using the *dc* component allows us to adjust the bias for both *amplitude* changes and *mode* changes, which is critical for single mode selection.

8.4.1.2 Amplifier operation

We now discuss how the nonlinear transfer function of the amplifier together with the movable bias scheme allows us to simultaneously meet the three stability requirements. Figure 8.7(a) shows the nonlinear transfer function of the amplifier. The transfer curve may be divided into the attenuation, gain, and voltage limiting regions. In the attenuation region the tangential slope at any point is less than 1 while in the gain region it is always greater than 1. In the voltage limiting region, the tangential slope is also less than 1 but was termed voltage limiting region due to the clipping of large input voltages that fall in the region.

Initially, the amplifier is biased at **A** in the gain region, which allows startup from ambient noise. As the oscillation grows and forms into pulses, the *dc* component of $V_Y(t)$ steadily increases. The amplifier uses

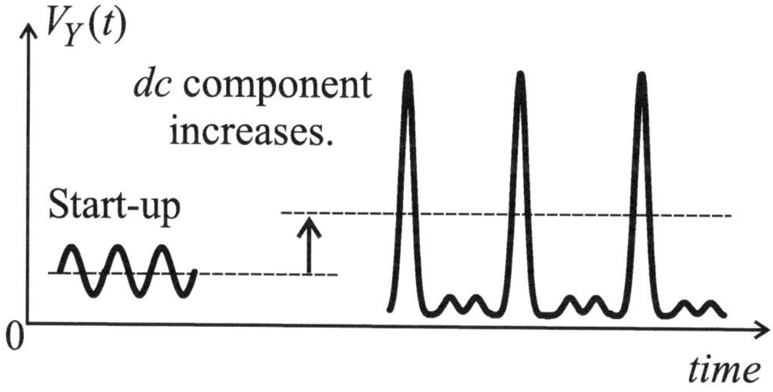

FIGURE 8.6
The *dc* component of V_Y increases as the oscillation grows and forms into a soliton pulse train. (From D. S. Ricketts, X. Li and D. Ham, "Electrical soliton oscillator," *IEEE Trans. on Microwave Theory and Tech.*, vol. 54, no. 1, pp. 373-382, Jan. 2006 © IEEE 2006.)

this increase in the *dc* component to lower its bias point as indicated with the broken arrow along the nonlinear transfer curve in Fig. 8.7(b). The reduced bias corresponds to a net *overall* gain reduction, since a portion of the pulse enters the attenuation region thereby reducing the amount of the pulse that receives gain. The bias point keeps moving down on the transfer curve to the steady-state bias point **B** in the attenuation region, where the net overall gain of the amplifier becomes equal to the loss in the system. This adaptive bias control resolves all the three issues mentioned above.

8.4.2 Stability Mechanisms — Solution

8.4.2.1 Distortion reduction

In the steady-state shown in Fig. 8.7(b), the input bias has been sufficiently reduced so that the peak portions of the input and output pulses do not go into the voltage limiting region of the amplifier transfer curve, thereby preventing significant signal distortion.

8.4.2.2 Perturbation rejection

In the steady-state shown in Fig. 8.7(b), small perturbations at the input of the amplifier are attenuated at the output since they fall in the attenuation region of the transfer curve. This perturbation rejection is possible since the steady-state bias point **B** has been reduced into the

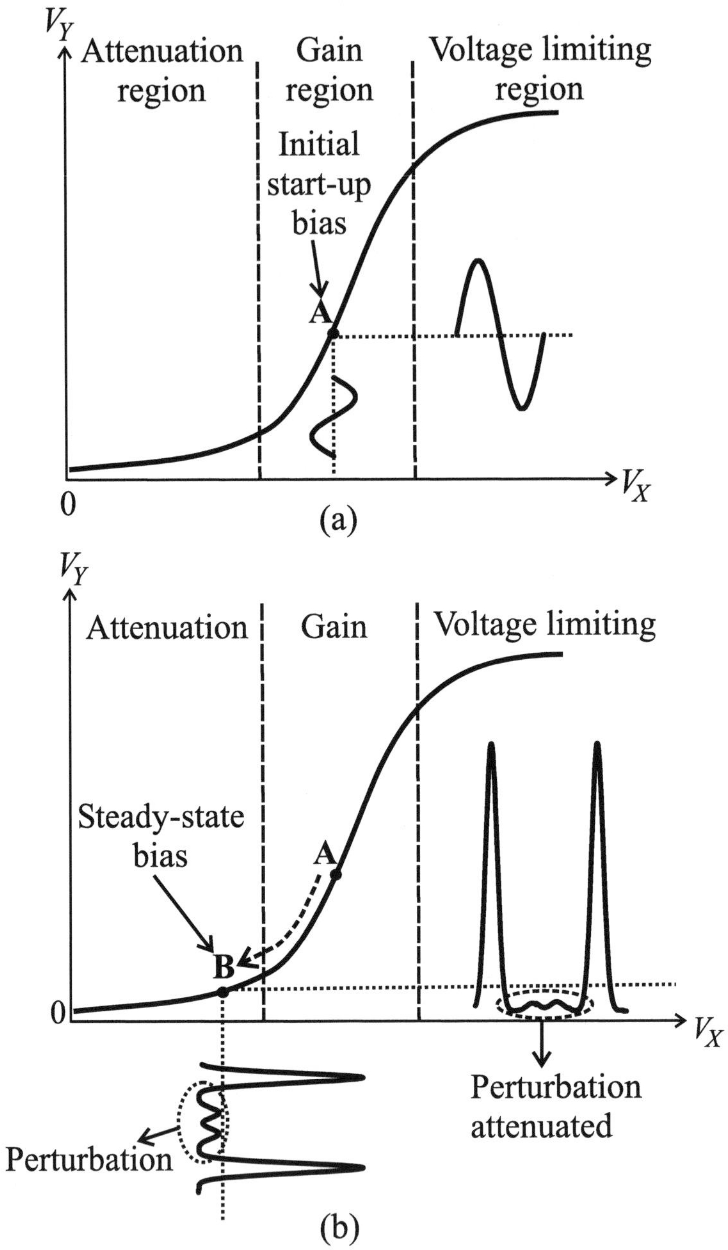

FIGURE 8.7
(a) Nonlinear transfer curve of our amplifier. In the initial startup, bias point **A** lies in the gain region. (b) The increased *dc* component lowers the bias point of the amplifier, leading to steady state bias point **B**.(From D. S. Ricketts, X. Li and D. Ham, "Electrical soliton oscillator," *IEEE Trans. on Microwave Theory and Tech.*, vol. 54, no. 1, pp. 373-382, Jan. 2006 © IEEE 2006.)

attenuation region. The higher portions of the pulses lying in the gain region still receive enough gain to compensate loss, however.

This threshold-dependent gain-attenuation mechanism vital for stabilizing our soliton oscillator has actually been widely employed in optical mode-locked systems where it is known as *saturable absorption* [24, 45]. The saturable absorption technique has seen limited use in the electrical domain, and there only for linear-pulse oscillators [22, 64].

8.4.2.3 Single mode selection

The single mode selection is also made possible by the adaptive bias control. Since a higher mode has a higher *dc* component and correspondingly a lower steady-state bias due to the adaptive bias control [Fig. 8.8], the higher mode receives a lower gain. One can take advantage of this mode-dependent gain to select a particular mode. Only those modes with sufficient gain to overcome the loss of the system can be sustained in steady-state oscillations. When more than one mode has sufficient gain, only the highest mode is stable since any small perturbation to a lower mode will grow into a soliton, resulting in a higher mode oscillation. Consequently, the mode-dependent gain allows only one soliton pulse train mode.

8.5 System Design and Amplifier Dynamics

In Subsec. 8.4.1.2 we presented a solution for stablizing solitons on a circular NLTL using an adaptive bias control to create the three stabalizing mechanisms described in Subsec. 8.4.2. The dynamics of the implementation of the adaptive bias are not trivial, however, and without consideration can lead to a system (i.e., not related to soliton dynamics) instability known as Q-switching[1].

Figure 8.5 shows Q-switching in a non-tuned version of the bipolar prototype described in Chapter 9. The oscillation starts up, but then dies suddenly, only to repeat this "start-up and die" process over and

[1]The term "Q-switching" is borrowed from the optical domain where the phenomenon of Q-switching is seen in many different systems. In its basic form, Q represents the quality factor of a cavity or resonator. Q-switching means modifying or "switching" the quality factor. This is normally done by reducing/increasing loss in the system. In our system, the loss is modulated by the amplitude of the oscillation via the saturable absorption mechanism; hence the instability comes from the loss, which in terms of a resonator can be thought of as the Q, of our oscillator switching from one value to another.

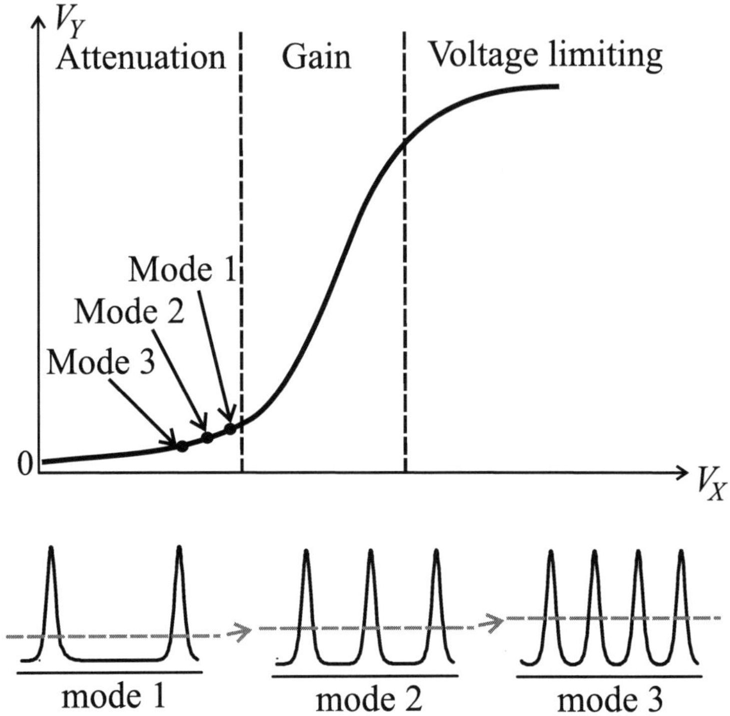

FIGURE 8.8
Mode-dependent steady-state bias.

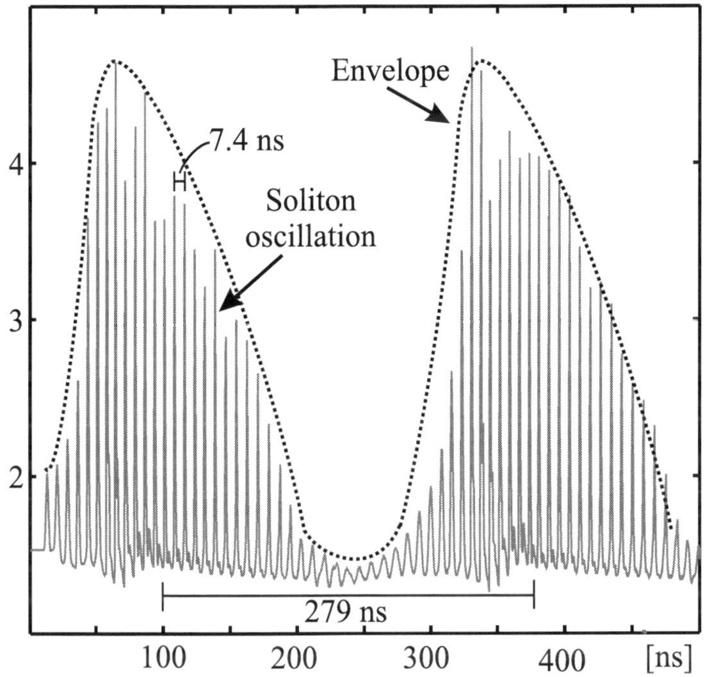

FIGURE 8.9
Q-switched instability. The soliton oscillation continuously starts up and then immediately dies again.

over again. Note that the soliton *pulse* repetition rate is 135 MHz (7.4 ns period), while the repetition rate of the Q-switching is approximately 3.6 MHz (279 ns period), an order of magnitude lower.

The Q-switched instability is caused by the underdamped response of the adaptive bias (Subsec. 9.1.2.1) being excited by the saturable absorber mechanism. To understand this phenomenon we must examine the dynamics of the amplifier and its effect on the overall system.

We start with the dynamics of the automatic bias control alone; we ignore the effects of the saturable absorption for the moment.

The dynamics of the adaptive bias control can be modeled as a system of two first-order differential equations. Since the adaptive bias looks at the average signal of the oscillation, we ignore the intra-cycle waveform and only consider the envelope of the oscillation (Fig. 8.9), which we label Λ. We also use the convention $\dot{x} = \frac{\partial x}{\partial t}$.

$$\dot{\Lambda} = \frac{(g-l)\cdot\Lambda}{T_R} \tag{8.1}$$

$$\dot{g} = \frac{(g_{nom}-\alpha\cdot\Lambda)-g}{\tau} \tag{8.2}$$

Equation (8.1) describes the change in amplitude of the pulse envelope, Λ, due to the excess gain, $g-l$, where g is the gain per cycle and l is the average loss in the system, per cycle[2]. T_R is the roundtrip time, or the inverse of the oscillation frequency, of the circular NLTL and sets the time scale for gain and loss, i.e., the pulse receives gain every T_R seconds and thus grows at that rate. Equation (8.2) describes the change in gain due to changes in the amplitude, Λ. With $\Lambda = 0$ the nominal gain of the amplifier is g_{nom}. The proportionality between the change of gain and the amplitude is modeled by a parameter α. τ represents the time constant of the adaptive bias, i.e., approximately the inverse of the LPF pole. The steady state solution for amplitude and gain can be easily found from (8.1) and (8.2)

$$\begin{aligned} 0 &= (g-l) \\ 0 &= (g_{nom}-\alpha\cdot\Lambda)-g \end{aligned}$$

or

$$\begin{aligned} g_s &= l \\ \Lambda_s &= \frac{g_{nom}-g_s}{\alpha}, \end{aligned}$$

where Λ_s and g_s are the steady state values.

This system can produces an under-damped response at start-up, Fig. 8.10. To understand this, let us examine a small perturbation, $\Delta\Lambda$ and Δg to the system in steady state, Λ_s and g_s:

$$\Lambda = \Lambda_s + \Delta\Lambda \tag{8.3}$$

$$g = g_s + \Delta g. \tag{8.4}$$

[2]Gain and loss are defined differently here than in a standard electronic circuit, where gain is typically defined as $g = \frac{V_{out}}{V_{in}}$. Here gain is defined as: *relative change in the output per unit change in time*, or in mathematical terms $g = \frac{\Delta V}{\Delta t} = \frac{V_{out}-V_{in}}{\Delta t}\frac{1}{V_{in}}$, where $\frac{1}{V_{in}}$ normalizes the gain with respect to the input. The result is that $g\cdot V_{in} = \Delta V = V_{out} - V_{in}$ in contrast to the circuit definition, $g\cdot V_{in} = V_{out}$.

Inserting these into (8.1) and (8.2), dropping the Δ^2 term and noting that the steady state terms are zero, i.e., $g_s - l = 0$, we obtain

$$\Delta\dot{\Lambda} = \frac{\Delta g \cdot \Lambda}{T_R} \tag{8.5}$$

$$\dot{g} = \frac{(\alpha \cdot \Delta\Lambda) - \Delta g}{\tau}. \tag{8.6}$$

Assuming a solution of the form

$$\begin{bmatrix} \Delta\Lambda \\ \Delta g \end{bmatrix} = \begin{bmatrix} \Delta\Lambda_o \\ \Delta g_o \end{bmatrix} \cdot e^{s \cdot t} \tag{8.7}$$

we obtain

$$\begin{bmatrix} s & \frac{\Lambda_s}{T_R} \\ \frac{\alpha}{\tau} & s + \frac{1}{\tau} \end{bmatrix} \begin{bmatrix} \Delta\Lambda_o \\ \Delta g_o \end{bmatrix} = 0. \tag{8.8}$$

A non-trivial solution exists if the determinant of the coefficients is zero,

$$s(s + \frac{1}{\tau}) + \frac{\Lambda_s}{T_R}\frac{\alpha}{\tau} = 0$$

$$s^2 + \frac{1}{\tau}s + \frac{\Lambda_s * \alpha}{T_R \tau} = 0 \tag{8.9}$$

resulting in

$$s = -\frac{1}{\tau} \pm \frac{1}{2}\sqrt{\frac{1}{\tau^2} - 4\frac{\Lambda_s \alpha}{T_R \tau}}. \tag{8.10}$$

We notice first that the real part of s is negative, so that any perturbation will always decay. Secondly, the system will will be damped whenever the imaginary part is zero,

$$\textbf{critically damped: } \tau < \frac{T_R}{4\Lambda_s \alpha}. \tag{8.11}$$

Physically this means that if the response time of the bias adjustment is fast enough, it can adjust for changes in amplitude (time scale T_R). If the bias adjustment is too slow, it will oscillate out of phase with the amplitude as it moves to a steady state value. A lower limit on τ is set

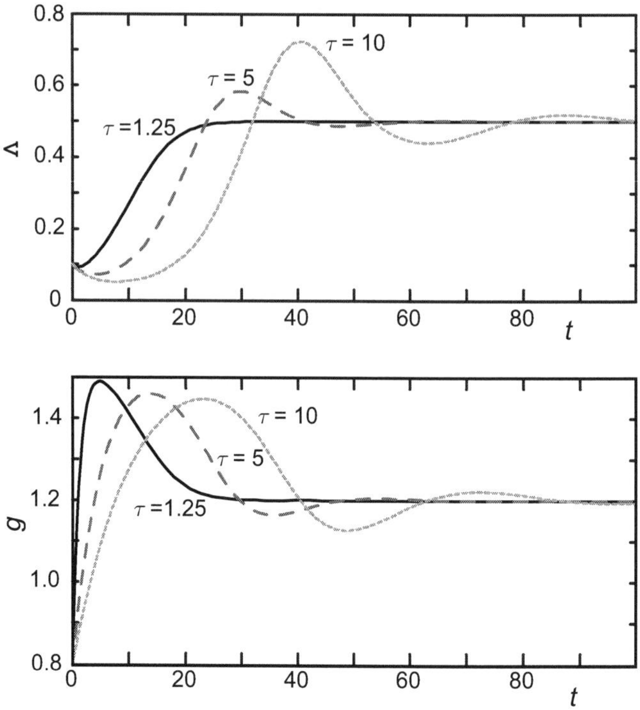

FIGURE 8.10
Simulation of the under-damped response of the adaptive bias. In this second order system we always reach a stable steady state after a short period of damped oscillation. $\tau = 1.25, 5, 10$, $l = 1.2$, $g_{nom} = 1.6$, $T_R = 2$ and $\alpha = .8$

by the fact that it needs to filter the individual pulses to find the *dc* average of the output waveform. Figure 8.10 shows simulation results of the adaptive bias dynamics for three values of τ. The amplitude, Λ, shows no oscillation when the condition (8.11) is met ($\tau = 1.25$). In all cases, however, the system eventually settles to a steady state amplitude and gain (remember the plot is of the oscillation *envelope*).

The Q-switched instability comes about when we introduce the saturable absorber mechanism. This seems counterintuitive, as the purpose of the saturable absorber was to attenuate perturbations and stabilize the oscillation. It does indeed do this; however, in our discussions we only considered the case for small perturbations *between* steady-state pulses. When we consider perturbations to the *amplitude* of the steady-state pulse stream we find that the saturable absorber excites the under-

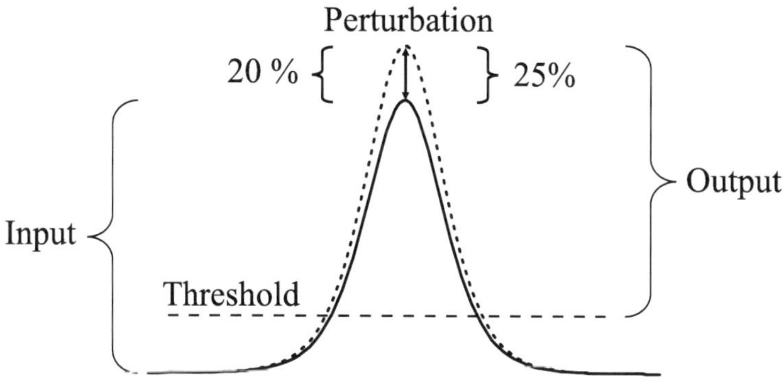

FIGURE 8.11
Saturable absorber instability. Perturbations to the input to the saturable absorber are *"amplified,"* a 20% increase in pulse input results in a 25% increase in pulse output.

damped response of the adaptive bias control, causing Q-switching of the oscillator (Fig. 8.9). To understand this "excitation," let us first look at the effect of a change in pulse amplitude for a **fixed** bias point. Figure 8.11 shows an example input pulse (solid line) whose magnitude has been increased by 20% (dashed line). This increase in input amplitude results in a 25% increase in the output of the saturable absorber. The additional 5% comes from the fact that at the output, the absolute value of the perturbation is compared to only the portion of the pulse above the gain-attenuation threshold, not the original input pulse amplitude and hence, the perturbation represents a larger portion of the signal. The result is that an instantaneous increase in amplitude will result in a larger gain through the amplifier [amplifier output = (saturable absorber output) · (amplifier gain)]. If the pulse were to reduce in amplitude, the opposite would occur and the pulse would receive less overall gain through the amplifier. Note that the increase from 20% to 25% was a result of the relative level of the gain-attenuation threshold; lower thresholds will result in a smaller increase, while higher thresholds a greater increase.

We can model the effects of the saturable absorber by replacing g with $g(\Lambda)$ or alternatively we can view this as an amplitude dependent loss, $l(\Lambda)$ [$g(\Lambda) \uparrow \Lambda \uparrow$ or $l(\Lambda) \downarrow \Lambda \uparrow$]. This latter viewpoint makes physical sense, as the saturable absorber is attenuating *less* of the pulse for an increase in amplitude.

We introduce a mathematically simple form of amplitude dependent loss, which models a reduction in loss as the amplitude increases:

$$l(\Lambda) = \frac{l_0}{1+\beta\cdot\Lambda} \tag{8.12}$$

where l_0 is the loss at zero amplitude and β is a proportionality constant.

Introducing $l(\Lambda)$ into (8.1) we can combine the effects of the adaptive bias and the saturable absorber amplitude dependent loss,

$$\dot{\Lambda} = \frac{1}{T_R}\left(g - \frac{l_0}{1+\beta\cdot\Lambda}\right)\cdot\Lambda \tag{8.13}$$

$$\dot{g} = \frac{(g_{nom} - \alpha\cdot\Lambda) - g}{\tau}. \tag{8.14}$$

The steady state values are found by setting $\dot{\Lambda} = 0$ and $\dot{g} = 0$

$$g_s = \frac{l_0}{1+\beta\cdot\Lambda_s}$$

$$\Lambda_s = \frac{g_{nom} - g_s}{\alpha}.$$

We now follow the same procedure as above, examining the response to a small perturbation around a steady state solution. To do this, we linearize the amplitude-dependent loss around $\Lambda = \Lambda_s$ as

$$l(\Lambda_s + \Delta\Lambda) \approx \frac{l_0}{1+\beta\cdot\Lambda_s} + \left.\frac{\partial l(\Lambda)}{\partial\Lambda}\right|_{\Lambda_s}\cdot\Delta\Lambda$$

$$l(\Lambda_s + \Delta\Lambda) \approx l_s - \frac{l_0\beta}{(1+\beta\cdot\Lambda_s)^2}\cdot\Delta\Lambda$$

$$l(\Lambda_s + \Delta\Lambda) \approx l_s - \gamma_{SA}\cdot\Delta\Lambda \tag{8.15}$$

where we have defined γ_{SA} as the second term to ease notation later. It is important to note that the loss *decreases* with an increasing amplitude, thus perturbations have a tendency to grow if the gain can not compensate by reducing. We now follow the same procedure above, ignoring Δ^2 terms and steady state terms, to obtain

$$\begin{bmatrix} s - \frac{\gamma_{SA}\Lambda_s}{T_R} & \frac{\Lambda_s}{T_R} \\ \frac{\alpha}{\tau} & s + \frac{1}{\tau} \end{bmatrix} \begin{bmatrix} \Delta\Lambda_o \\ \Delta g_o \end{bmatrix} = 0. \tag{8.16}$$

yielding

$$s^2 + \left(\frac{1}{\tau} - \frac{\gamma_{SA}\Lambda_s}{T_R}\right) \cdot s + \left(-\frac{\gamma_{SA}\Lambda_s}{T_R\tau} + \frac{\Lambda_s * \alpha}{T_R\tau}\right) = 0 \tag{8.17}$$

or

$$s = -\frac{1}{2}\left(\frac{1}{\tau} - \frac{\gamma_{SA}\Lambda_s}{T_R}\right) \pm \sqrt{\left(\frac{1}{\tau} - \frac{\gamma_{SA}\Lambda_s}{T_R}\right)^2 - 4\left(-\frac{\gamma_{SA}\Lambda_s}{T_R} + \frac{\Lambda_s * \alpha}{T_R\tau}\right)} \tag{8.18}$$

We can see immediately that the solution has significantly changed. Without the saturable absorber, the real part was always negative. In (8.18) the real part can be positive. For values of τ and Λ, such that the square root is always imaginary (this is the case for typical values) the condition for stability, i.e., decaying response, can be stated as

$$\textbf{stability criteria: } \tau < \frac{T_R}{\gamma_{SA}\Lambda_s}. \tag{8.19}$$

Even if the real part is negative, the response can still exhibit oscillations when s contains an imaginary part.

$$\textbf{critically damped: } 4\left(-\frac{\gamma_{SA}\Lambda_s}{T_R} + \frac{\Lambda_s * \alpha}{T_R\tau}\right) < \left(\frac{1}{\tau} - \frac{\gamma_{SA}\Lambda_s}{T_R}\right)^2, \tag{8.20}$$

Figure 8.12(a) shows the system response for values of τ below and above the critical damped $\tau = 1.2$. We can easily see that for τ above 1.2, the system begins to show an underdamped response. Figure 8.12(b) shows the system response for values of τ below and above the critical stabile $\tau = 65$. When τ is below this threshold, initial perturbations die out in an underdamped response. Above this value, the system begins to oscillate and the amplitude grows, demonstrating the origin of the Q-switching instability.

We can understand qualitatively what is happening by examining Fig. 8.13, which shows a close-up of the waveform for large instabilities, i.e., Q-switching as seen in Fig. 8.9. The oscillation envelope (solid line), which represents the amplitude of the solitons oscillating in the loop, rises rapidly only to die away a short time later. The gain of the amplifier (dashed line) moves in response to the rise and fall of the amplitude, but is not able to stabilize the amplitude of the oscillation.

The process can be understood as follows. At start-up there is excess gain and the oscillation envelope starts to rise. As the amplitude

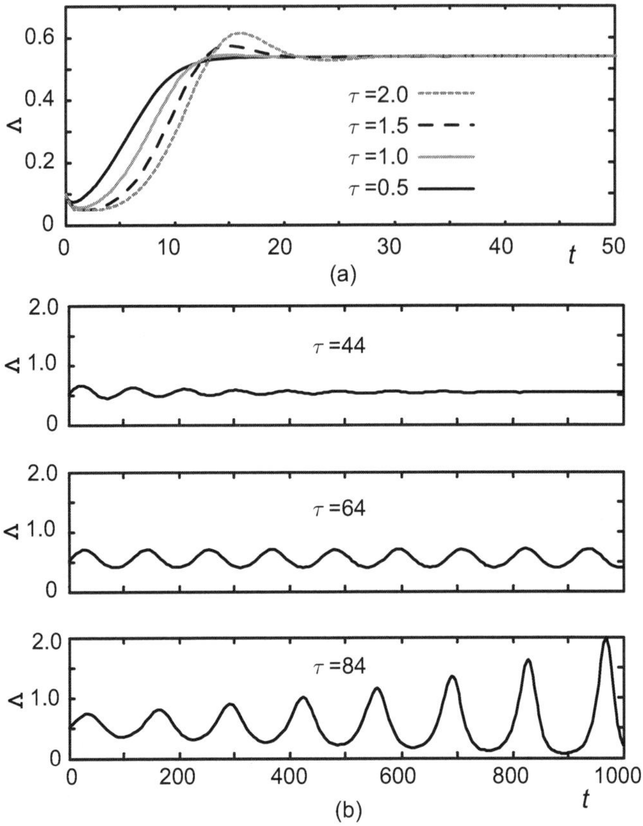

FIGURE 8.12
Simulation of the stable and unstable response of the adaptive bias with saturable absorption. (a) τ near critically damped value, $\tau_c = 1.2$. (b) τ near the critically stable value, $\tau = 65$. $g_{nom} = 1.6$, $T_R = 2$, $\beta = .05$, $l_0 = 1.2$ and $\alpha = .8$

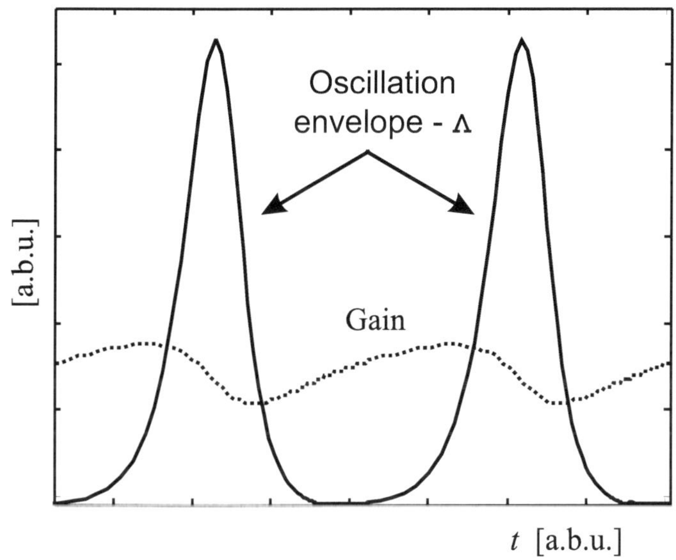

FIGURE 8.13
Behavioral simulation of Q-switching instability. The oscillation **envelope** rises and then suddenly dies out. This process repeats ad infinitum. The gain rises and falls in response to the amplitude change. $\tau = 50$, $l = 1.2$, $g_{nom} = 1.5$, $\alpha = .8$, and $\beta = .05$

increases, the loss will reduce due to the amplitude dependent loss in the saturable absorber ($\Lambda \uparrow \ l(\Lambda) \downarrow$), causing the pulse train to grow even faster. Eventually, the adaptive bias is able to catch up and arrest this increasing amplitude by lowering the gain of the amplifier. As soon as the pulses start to reduce in amplitude, however, the loss increases ($\Lambda \downarrow \ l(\Lambda) \uparrow$), driving the envelope of the pulse amplitude to zero. The gain then increases to startup the oscillation again and the process repeats itself.

The solution for eliminating the Q-switched instability is to increase the speed of the adaptive bias such that it can compensate for the increase/decrease in loss due to the changes of pulse amplitude. This was shown in Fig. 8.12.

8.6 Summary

With the development of the three soliton stabilizing mechanisms and correct amplifier/SA gain and time constants, we can now proceed to construct a prototype of the electrical soliton oscillator. The concepts presented are general and in the following chapters we will implement them in three different prototypes. This will unequivocally demonstrate the validity of these methods as well as demonstrate their practical implementation.

9

The Circular Soliton Oscillator

CONTENTS

In this chapter we describe the implementation of the soliton oscillator concept introduced in Chapter 8 in the form of three prototypes: a low MHz, discrete CMOS prototype; a microwave, discrete bipolar prototype; and a CMOS chip-scale prototype.

9.1 CMOS, Low MHz Prototype

The low-frequency operation and discrete nature of this prototype allow us to examine not only the stable soliton oscillation, but also the inner workings of the oscillator, including the key stabilizing mechanisms. We will first describe the design and fabrication of the low MHz prototype and then experimentally verify the key operating principles outlined in Sec. 8.4.

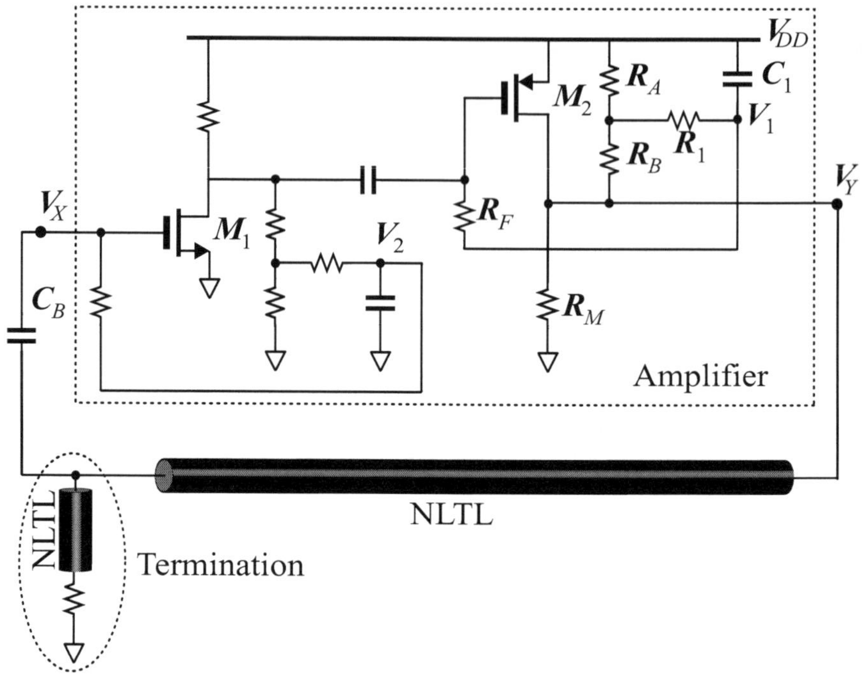

FIGURE 9.1
Detailed oscillator schematic. (From D. S. Ricketts, et al., "On the self-generation of electrical soliton pulses," *IEEE J. Solid State Circuits*, vol. 42, no. 8, pp. 1657-1668, Aug 2007 © IEEE 2007.)

9.1.1 Oscillator Implementation

Figure 9.1 shows the detailed schematic of the low MHz soliton oscillator prototype.

The amplifier was built using discrete CMOS transistors (Advanced Linear Devices: ALD1103PB) and is shown inside of the dashed box. The NLTL, shown at the bottom, was built from discrete inductors and *pn* junction varactors. Table 9.1 summarizes the components used and their key parameters: L, C_0, b, R_L and R_C, which were measured using an HP4275A LCR meter and an Agilent 4156C Semiconductor Analyzer. Note that the inductor has a frequency-dependent loss which is significant for frequencies above 1 MHz.

In addition to the amplifier and the NLTL there are two components in Fig. 9.1 that are not shown in the basic topology (Fig. 8.2): the

TABLE 9.1
Discrete NLTL components.

Component	Manufacturer	Part #	Parameter	Value
L	Panasonic	ELJEA220KF	L	23 μ H
L	Panasonic	ELJEA220KF	R_L	0.65 Ω (*dc*)
L	Panasonic	ELJEA220KF	ESR	2 Ω (1 MHz) 260 Ω (10 MHz)
C	Zetex	ZMV934TA	C_0	140 pF
C	Zetex	ZMV934TA	b	0.11
C	Zetex	ZMV934TA	R_C	1.48 Ω
C	Zetex	ZMV934TA	G	0 Ω

blocking capacitor, C_B, and the termination. The capacitor couples the voltage signal from the NLTL to the input of the amplifier. The termination is used to absorb the energy of the pulse once it has propagated down the NLTL. The termination of the NLTL is not trivial since its characteristic impedance varies with signal voltage $\left(z = \sqrt{\frac{L}{C(V)}}\right)$.

9.1.1.1 Amplifier design

The amplifier consists of two inverting stages, one built around an NMOS transistor, M1, and the other built around a PMOS transistor, M2, and together they form an overall non-inverting active network. The adaptive bias scheme is implemented for both inverting stages. It functions as follows for the PMOS stage. The output waveform, V_Y, is sensed by the voltage divider consisting of the two resistors, R_A and R_B, and then is integrated by the R_1-C_1 low pass filter (LPF). The integrated voltage, V_1, represents a scaled *dc* component of the waveform, V_Y. This *dc* component is fed back to the gate of M2 to set its bias. As the *dc* level of V_Y increases, V_1 will rise with respect to ground. The increase in V_1 corresponds to a reduction in the gate-source voltage of M2, effectively lowering its bias. A similar argument applies to the NMOS stage. Combining the two stages, the bias of the amplifier at the input is reduced as the *dc* component of V_Y increases, performing the adaptive bias control described in Subsec. 8.4.1.2.

The design of the amplifier is highly dependent on the NLTL used and its characteristics. With the aid of a circuit simulator, however, a stable oscillator can be easily designed. There are three main design points within the amplifier: the resistor divider ratio (R_A and R_B), the pole of the LPF (R_1 and C_1), and the gain per stage (transconductance,

g_M, of M and R_M). The resistor divider is set so that the average voltage in steady-state of the LPF, i.e., V_1 and V_2, is just above the transistor threshold (i.e., V_T); this allows the lower portion of the signal to fall below threshold and be attenuated (saturable absorption). The LPF pole is set low enough to average the signal, but high enough to allow the gain to respond to external perturbations. The critical frequencies are discussed in Chapter 8. The gain of the system is set such that during steady-state operation the gain exactly cancels the loss of the NLTL and the desired amplitude is achieved. The exact value can only be determined by running system simulations with a well modeled NLTL.

9.1.1.2 NLTL design

The nonlinear transmission line is the most important part of the oscillator design, as it determines the two key metrics of the oscillator: the pulse width and the frequency (the amplitude is set by the amplifier). Figure 9.2 depicts a steady state waveform of the soliton oscillator along with the NLTL and its design parameters, L, C_0, and b. The pulse width, w, and oscillating frequency, f, are given as:

$$w_{time} = \frac{3 - 2bA}{\sqrt{6bA}} \cdot \sqrt{LC_0} \tag{9.1}$$

$$f = \frac{1}{\sqrt{LC_0}} \cdot \sqrt{\frac{3}{3 - 2bA}} \cdot \frac{n}{N}, \tag{9.2}$$

where n is the order of the mode [See. (2.27)], and N is the total number of sections, or length of the transmission line. Hence, the pulse width can be exactly engineered from 1 second to 1 picosecond. In practice, however, the parasitics of the NLTL and the bandwidth of the amplifier will set the ultimate limit on the shortest pulse width. Typically below 100 ps, the NLTL is implemented as a coplanar strip line (CPS) periodically loaded with varactors in order to reduce parasitics.

9.1.1.3 Termination

In our prototype we used two types of termination. The first is a simple resistor with a value that is the average of the characteristic impedance seen by a desired signal. The second termination is a combination of a high-loss NLTL section terminated by a resistor, Fig. 9.3. The resistance is intentionally added at an increasing rate to "softly absorb" the pulse and reduce reflections (this method was adapted from [65]). There will still be reflections from the termination; however, they will be attenuated by the *saturable absorption* in the amplifier (See Subsec. 9.1.2.3).

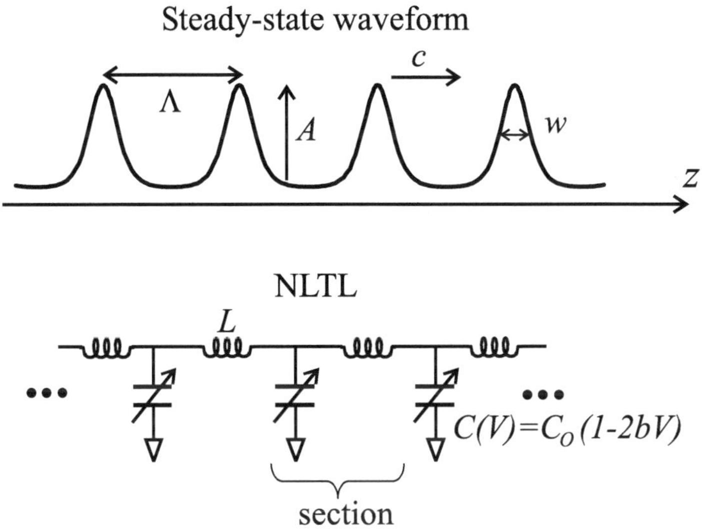

FIGURE 9.2
Depiction of the steady-state waveform of the soliton oscillator. The pulse width and oscillation frequency are set directly by the NLTL design parameters: L, C_0, and b.

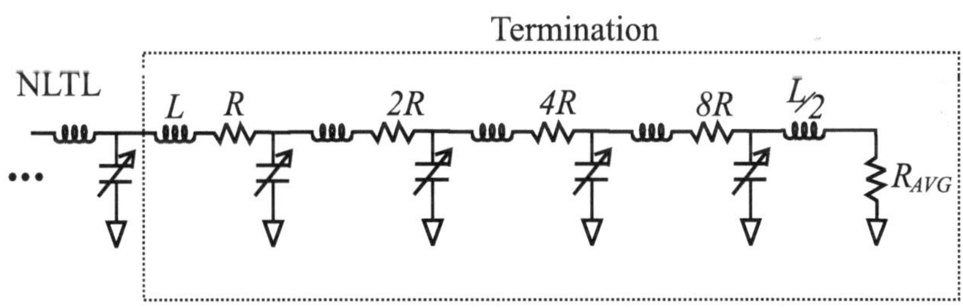

FIGURE 9.3
The termination for the NLTL is constructed from a high-loss section of NLTL where additional series resistance is added. The section of high-loss NLTL is terminated with a resistor, whose value is the average resistance of the NLTL seen by the steady-state waveform.

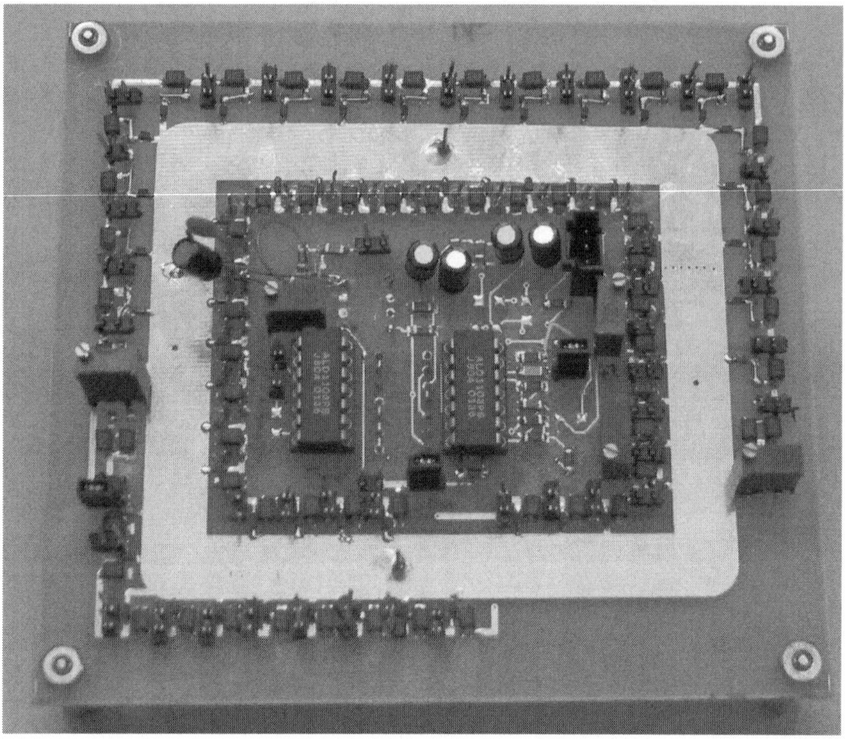

FIGURE 9.4
Low MHz soliton oscillator prototype. (From D. S. Ricketts, X. Li and D. Ham, "Electrical soliton oscillator," *IEEE Trans. on Microwave Theory and Tech.*, vol. 54, no. 1, pp. 373-382, Jan. 2006 © IEEE 2006.)

9.1.2 Experimental Verification

The low MHz prototype is shown in Fig. 9.4. It was designed for a repetition rate of around 1 MHz and a soliton pulse width of about 100 ns. The prototype produces a stable soliton pulse train after an initial pulse-forming transient process. In the following, we present a sequence of measurements that attest to the validity of the soliton oscillator concept and stability mechanisms.

9.1.2.1 Adaptive bias control

Figure 9.5(a) shows the voltage signal measured at the output of the amplifier during the oscillation startup transient while Fig. 9.5(b) is the corresponding, measured bias adjustment in the amplifier. As the oscillation grows and forms into soliton pulses, the amplifier self-adjusts to lower its bias according to the adaptive bias control scheme explained in Subsec. 8.4.1. The bias point eventually settles to **B**, at which time the net overall gain of the amplifier becomes equal to the system loss. Figure 8.7(b) is repeated in Fig. 9.5(c) for convenience, and hypothetically illustrates the bias adjustment in the transient process. Note that the bias adjustment exhibits an under-damped response. A more detailed discussion of the dynamics of the bias adjustment are described in Chapter 8.

9.1.2.2 Startup soliton dynamics

Figure 9.6 shows a detailed view of the oscillation startup measured after the 8th section on the NLTL (A total 22 *LC* sections were used in this specific experiment to best illustrate the startup dynamics). The oscillator starts by amplifying ambient noise creating a small oscillation, eventually growing into a steady-state soliton pulse train. During this process, another competing mode is clearly seen: it first grows with time, but is eventually suppressed by the stabilizing mechanism of the amplifier. In the figure, one can also observe that the shorter pulse (competing mode) propagates at a different speed than the taller pulse (main mode that survives): in this time-domain measurement at the fixed point on the NLTL, the shorter pulse originally behind the taller pulse catches up with the taller pulse and eventually moves ahead of it after collision (See Fig. 6.2). In the space domain, this corresponds to the taller pulse propagating faster than the shorter pulse, which is a key signature of solitons on the NLTL as explained in Chapter 2. The variation in pulse amplitude seen in Fig. 9.6 is a result of the transient under damped bias adjustment, similar to what is shown in Fig. 9.5(b). The waveform eventually reaches a stable oscillation with constant pulse amplitude and spacing as will be seen shortly in Subsec. 9.1.2.4.

9.1.2.3 Perturbation rejection

Figure 9.7 shows input-output measurements of the *stand-alone* amplifier when the amplifier is biased at its steady-state bias point **B**. The test input signal consists of two main pulses and perturbations between them. The perturbations are significantly attenuated at the output of the amplifier while the main pulses are amplified, due to the threshold-dependent gain-attenuation mechanism (saturable absorption, Subsec. 8.4.1). Note that in addition to the perturbation attenuation, the amplifier has sharpened the main pulses. This is because the saturable absorption mechanism attenuates the lower portion of the input main pulses while amplifying their higher portion.

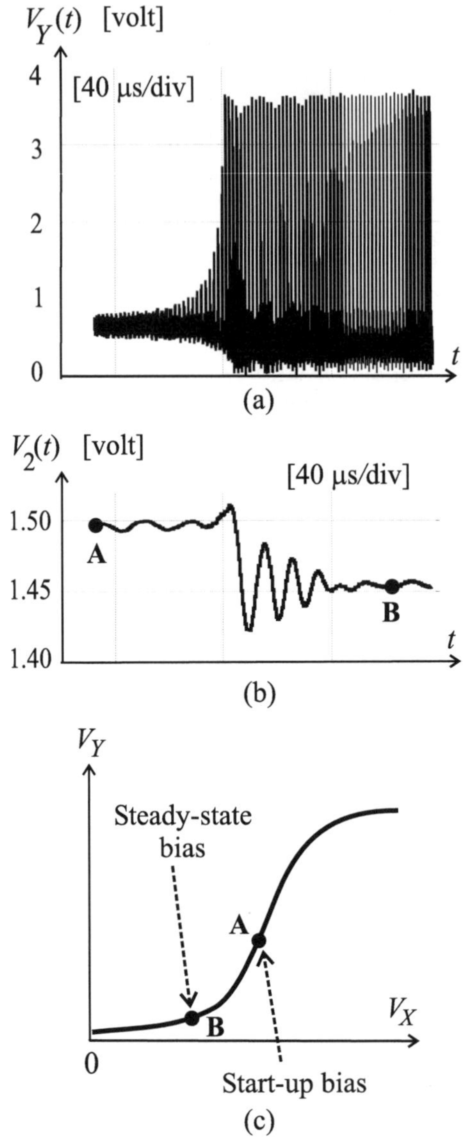

FIGURE 9.5
(a) Measured voltage signal at the output of the amplifier during the oscillation startup transient. (b) Measured bias response, $V_2(t)$, of the amplifier ($V_2(t)$ is with reference to Fig. 9.1) filtered by the oscilloscope with a 500 kHz BW. $V_2(t)$ and $V_X(t)$ in the amplifier of Fig. 9.1 have the same *dc* component. (c) Redrawing of Fig. 8.7(b). (From D. S. Ricketts, X. Li and D. Ham, "Electrical soliton oscillator," *IEEE Trans. on Microwave Theory and Tech.*, vol. 54, no. 1, pp. 373-382, Jan. 2006 © IEEE 2006.)

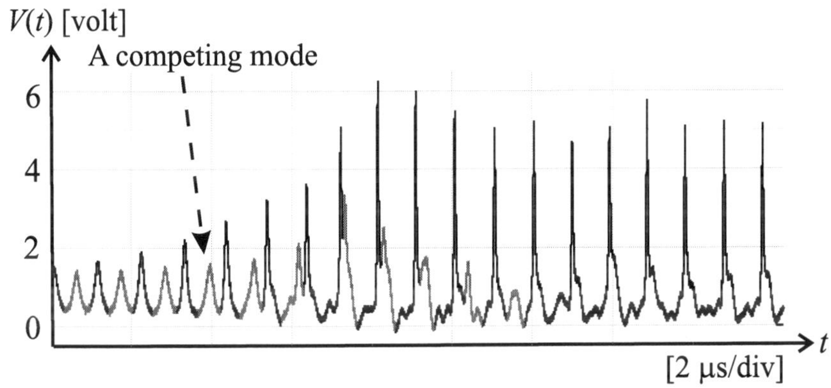

FIGURE 9.6
Measured startup transient. (From D. S. Ricketts, X. Li and D. Ham, "Electrical soliton oscillator," *IEEE Trans. on Microwave Theory and Tech.*, vol. 54, no. 1, pp. 373-382, Jan. 2006 © IEEE 2006.)

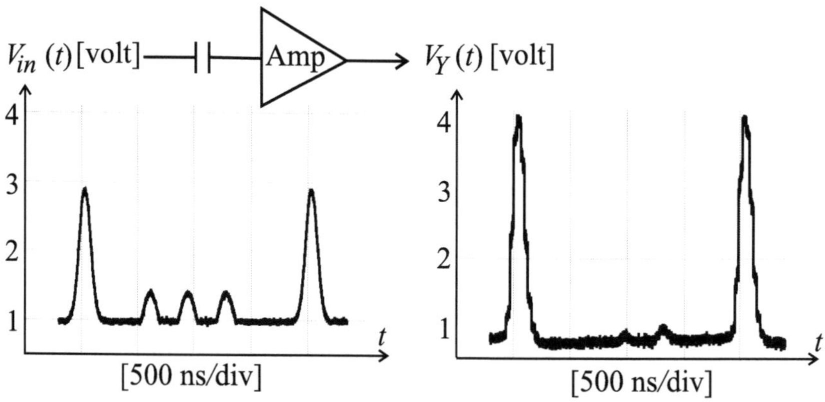

FIGURE 9.7
Input-output measurements of the stand-alone amplifier biased at **B**. Note that the line width variation in the output waveform is due to the resolution of the measurement system at this scale setting; by zooming in on a single pulse, the output has the same smoothness as the input. (From D. S. Ricketts, X. Li and D. Ham, "Electrical soliton oscillator," *IEEE Trans. on Microwave Theory and Tech.*, vol. 54, no. 1, pp. 373-382, Jan. 2006 © IEEE 2006.)

9.1.2.4 Steady-state soliton oscillation

Figure 9.8(a) shows a steady-state soliton pulse train measured after the 8th section on the NLTL (30 *LC* sections total). This waveform corresponds to the $l = L$ mode (Mode 1) shown in Fig. 8.1(b): at this mode, there exists only one pulse propagating around the NLTL. The measured period, T_{mode1}, for this particular mode was 1.3 μs.

By tuning the gain (bias), $l = 2L$ mode (Mode 2) oscillation shown in Fig. 9.8(b) was controllably obtained. This mode corresponds to the $l = 2L$ mode (Mode 2) shown in Fig. 8.1(b), and two pulses co-propagate in the NLTL. The measured period, T_{mode2}, for this mode was 0.7 μs. It is noteworthy that $2T_{mode2} \neq T_{mode1}$ while $2L_{mode2} = L_{mode1}$. This is because the two modes have different amplitudes, and hence propagate with different speeds, another signature of the soliton propagation.

In both oscillation modes, the amplitude and pulse repetition rate remained stable, showing no discernable variation. Even when deliberately perturbed with large external signals, Fig. 9.9, the oscillation always returned to the same steady-state soliton pulse train. Additionally, for a given set of circuit parameters, every start-up led to the same steady-state oscillation. These experiments demonstrate the level of robustness and controllability as found in traditional *LC* oscillators.

9.1.2.5 Soliton propagation in steady-state

Figure 9.10 shows the steady-state waveforms of the oscillator measured at three different *positions* on the NLTL, **A**, **B** and **C**. This experiment elucidates how the pulse shape changes as it circulates in the oscillator. The pulse at the output of the amplifier, **A**, is not an exact soliton and, hence, sharpens into a soliton propagating down the NLTL. Once the soliton is formed at **B**, it exhibits damping soliton dynamics as it further travels down the NLTL due to the loss (**B** $\rightarrow$ **C**), lowering in amplitude and velocity while increasing in width, as explained in Sec. 5.5.

It is this clear existence of the transition point, **B**, between the pulse sharpening and the stable damping that indicates the formation of the soliton at that transition point. This may be seen more clearly by plotting the product of the pulse width and the square root of amplitude as shown in Fig. 9.11. Each process, soliton formation and soliton damping, has a distinctly different trajectory and hence, are distinct separate processes. Moreover, the solid line in Fig. 9.11 is not simply a curve fit, but rather is the theoretical prediction of the amplitude-width-product for an ideal soliton using the *measured* NLTL parameters and (9.1):

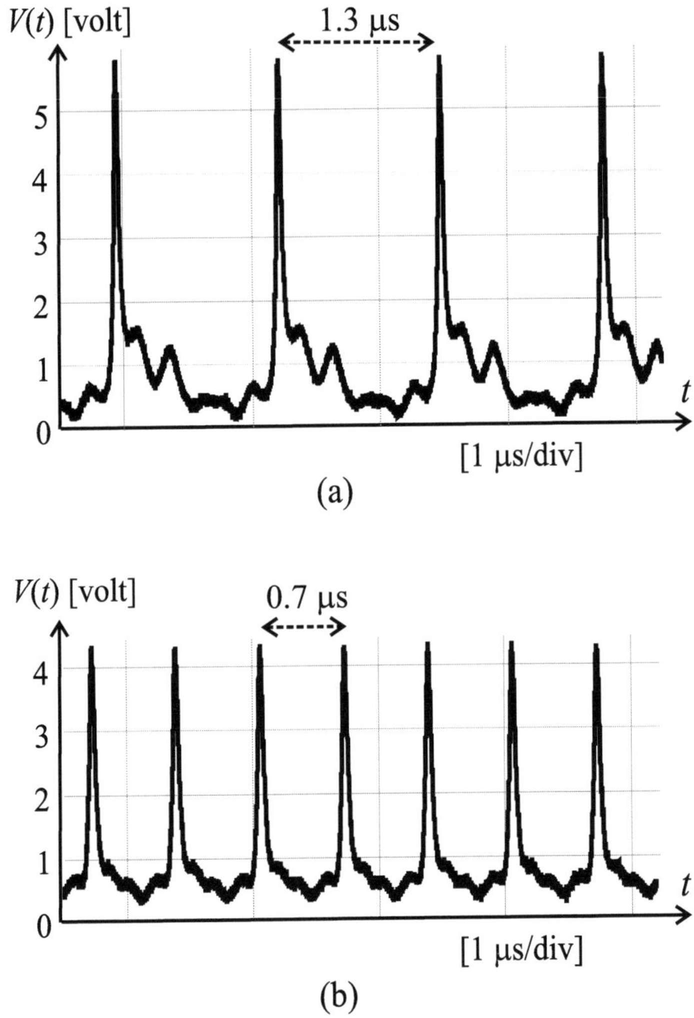

FIGURE 9.8
Measured stable soliton oscillation. (a) $l = L$ oscillation. (b) $l = 2L$ oscillation. (From D. S. Ricketts, X. Li and D. Ham, "Electrical soliton oscillator," *IEEE Trans. on Microwave Theory and Tech.*, vol. 54, no. 1, pp. 373-382, Jan. 2006 © IEEE 2006.)

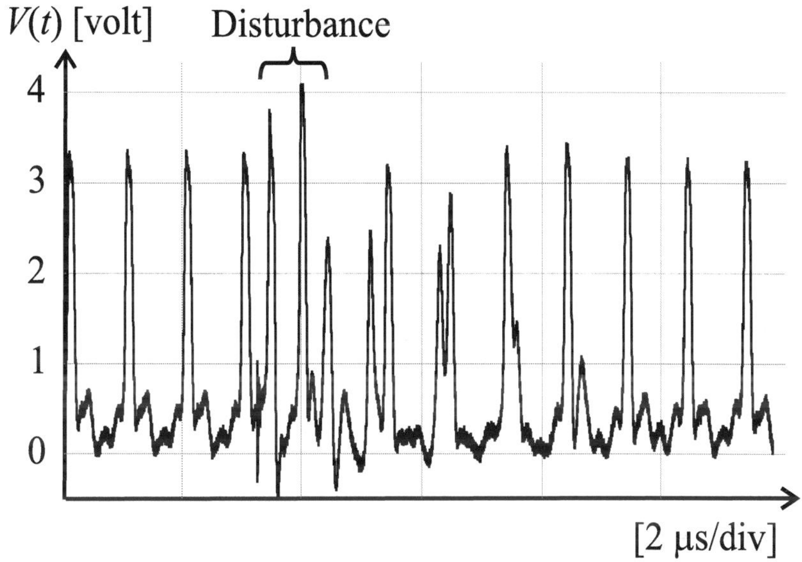

FIGURE 9.9
Recovery of steady-state oscillation to a large signal disturbance. Two additional pulses were artificially added by capacitively coupling a signal onto the NLTL. The oscillation recovers in just a few cycles.

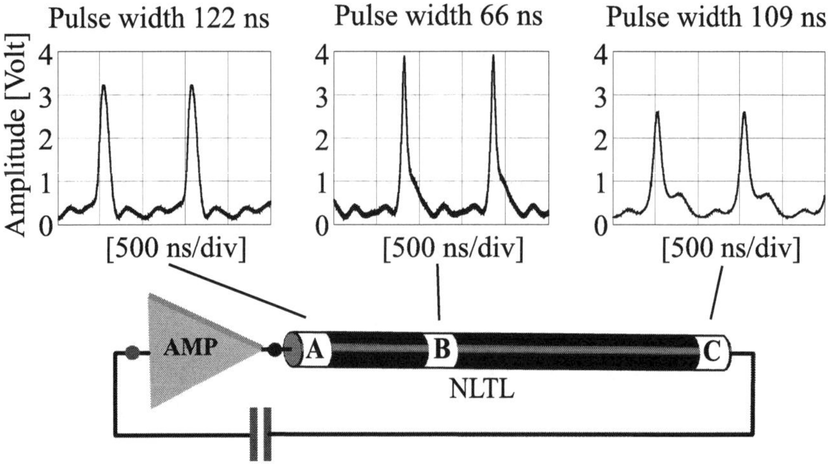

FIGURE 9.10
Experimental results of the steady-state oscillation at three points, **A**, **B** and **C**, along the NLTL.

$$
\begin{aligned}
w_{time} \cdot \sqrt{A} &= \frac{3 - 2bA}{\sqrt{6bA}} \cdot \sqrt{LC_0} \cdot \sqrt{A} \\
&= \left(-\frac{2\sqrt{b}}{\sqrt{6}} \sqrt{LC_0} \right) \cdot A + \frac{3}{\sqrt{6b}} \sqrt{LC_0}, \qquad (9.3)
\end{aligned}
$$

9.2 Bipolar, Microwave Prototype

Due to the low transconductance of CMOS devices, the effective saturable absorbtion is not large. Bipolar technology offers a significantly higher transconductance and with it a much larger saturable absorbtion. This section describes a bipolar implementation of the circular soliton oscillator operating in the microwave region. The oscillator follows the CMOS design in concept, but uses a different amplifier implementation where the gain and saturable absorbtion are concentrated into a single stage. In addition, the simplicity of this amplifier design allows an examination of the amplifier dynamics, in particular the adaptive bias control and saturable absorption response times.

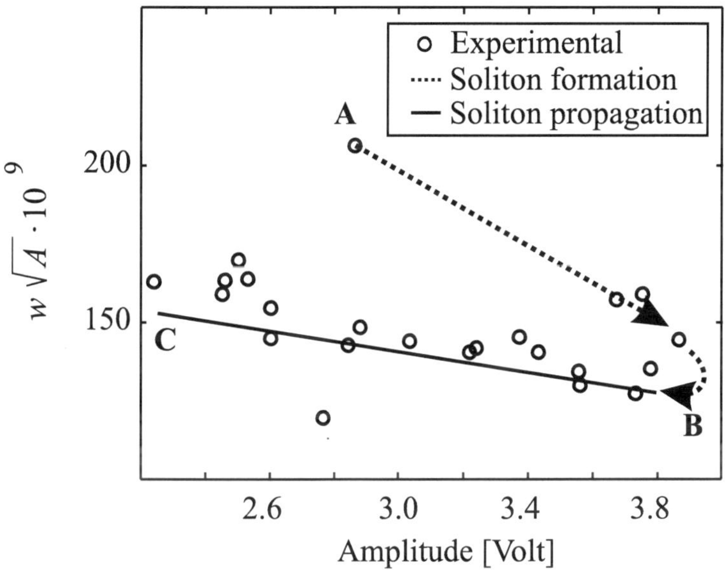

FIGURE 9.11
Pulse-width-square-root-amplitude product. Pulse sharpening from **A** to **B** is shown with the dashed line. Soliton propagation from **B** to **C** is shown with the solid line. The intersection at point **B** of the two distinct trajectories demonstrates clearly that a soliton is formed at **B**. (From D. S. Ricketts, et al., "On the self-generation of electrical soliton pulses," *IEEE J. Solid State Circuits*, vol. 42, no. 8, pp. 1657-1668, Aug 2007 © IEEE 2007.)

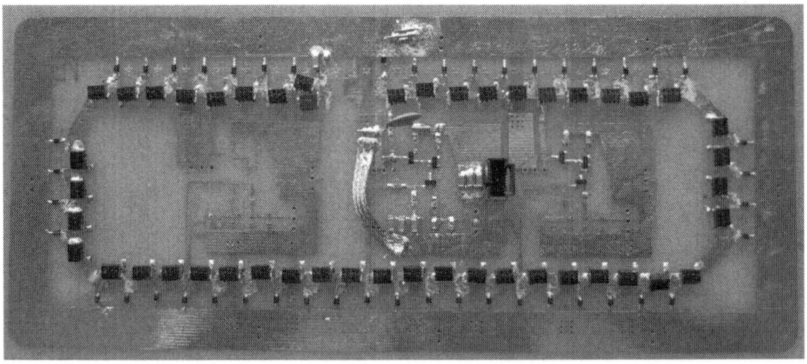

FIGURE 9.12
Microwave soliton oscillator prototype. (From D. S. Ricketts, X. Li and D. Ham, "Electrical soliton oscillator," *IEEE Trans. on Microwave Theory and Tech.*, vol. 54, no. 1, pp. 373-382, Jan. 2006 © IEEE 2006.)

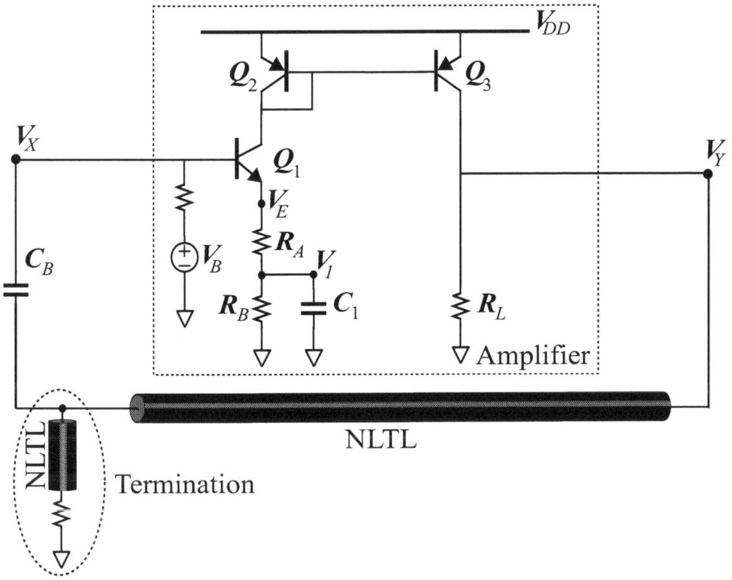

FIGURE 9.13
Microwave soliton oscillator schematic. (From D. S. Ricketts, et al., "On the self-generation of electrical soliton pulses," *IEEE J. Solid State Circuits*, vol. 42, no. 8, pp. 1657-1668, Aug 2007 © IEEE 2007.)

9.2.1 Oscillator Implementation

The microwave soliton oscillator prototype was implemented using a discrete NLTL, similar to the low MHz prototype, and an amplifier built from RF bipolar transistors (Philips BFT2 and BFT3). Figure 9.13 shows the detailed schematic of the oscillator. The NLTL was constructed from Zetex varactors (ZMV930; $\sim$ 5 pF nominal capacitance) and Coilcraft RF inductors (B07T-L-22NG-LB; 22 nH). The termination was a simple resistor, whose value is the average impedance of the NLTL.

The amplifier, shown in Fig. 9.13 inside the dashed box, consists of a common emitter amplifier with emitter degeneration and a signal inversion stage. The amplifier contains the same functionality as the CMOS amplifier, although its detailed operation differs. The biasing for the amplifier consists of three components: R_A, R_B, and C_1. At *dc*, C_1 is an open circuit and the bias of the amplifier is set by V_B and R_A+R_B. During *ac* operation, however, C_1 shorts R_B (to first order) and the *ac* gain is simply: $g_{ac} = \frac{V_X}{R_A} \cdot R_L$. The non-inverting function is obtained by mirroring the output current of Q_1 through the *PNP* current mirror into R_L, such that an increase in V_X corresponds to an increase in V_Y.

The adaptive bias adjustment in the microwave amplifier achieves the same functionality as the CMOS amplifier; however, its operation is significantly different. Instead of the adaptive bias moving the bias point, **B** in Fig. 8.7, down the transfer curve, the transfer curve is moved instead, such that V_X lies at the same *relative* position on the transfer curve as **B** in Fig. 8.7. This is shown qualitatively in Fig. 9.14.

The threshold between gain and attenuation is effectively set by the *dc* voltage on V_1. When the input, V_X, is above $V_1 + V_{BE}$, the amplifier provides gain: $g_{ac} = \frac{V_X}{R_A} \cdot R_L$. When it is below $V_1 + V_{BE}$, Q_1 turns off and the gain drops to zero (attenuation). Thus a very strong threshold-dependent gain is achieved. The adjustment of V_1 is done with the voltage divider, R_A and R_B. It should be noted that the transition between gain and attenuation is significantly sharper than for the CMOS amplifier; this made it easier to use a single saturable absorber in the microwave prototype (the CMOS prototype had two saturable absorbers, one in each stage of the amplifier).

Simulation results demonstrating the operation of the "bias" adjustment process are shown in Figs 9.15 and 9.16. Figure 9.15 shows the start-up transient of the oscillator (simulation). The input minus one V_{BE}, $V_X - V_{BE}$, is shown in gray; its *dc* bias point does not change during the transient. The voltage V_1, on the other hand, rises during transient and settles to a steady-state value. At this steady-state point, the lower portions of the input are attenuated and the higher portions are amplified. Figure 9.16(a) shows a close-up during steady-state. The biasing of the input, $V_X - V_{BE}$, across V_1 is clearly seen. Figure 9.16(b) shows the output, V_Y, of the amplifier. The dashed lines connect the output of the amplifier to the input; we can see clearly that only the portion above V_1 is amplified. This figure shows in the clearest terms the operation of the saturable absorber function in the soliton oscillator. In addition, we can also see explicitly how the input pulse is sharpened during amplification, a secondary effect of saturable absorption.

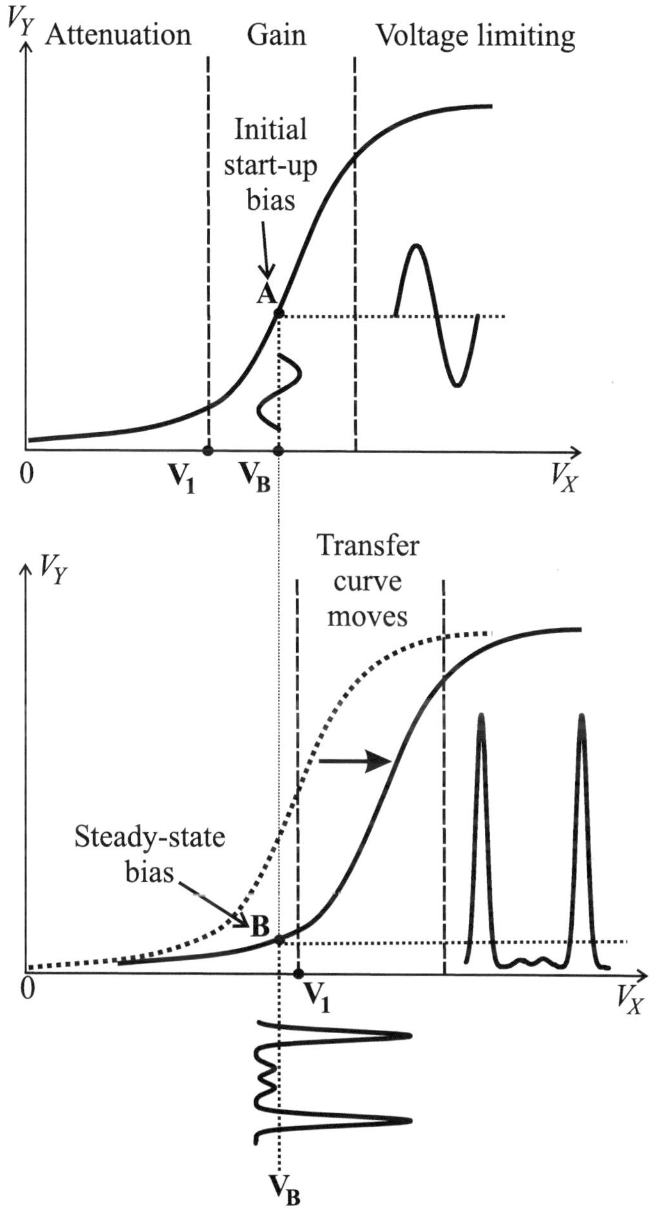

FIGURE 9.14
Amplifier operating principle. Transfer curve moves up instead of the bias moving down. (From D. S. Ricketts, et al., "On the self-generation of electrical soliton pulses," *IEEE J. Solid State Circuits*, vol. 42, no. 8, pp. 1657-1668, Aug 2007 © IEEE 2007.)

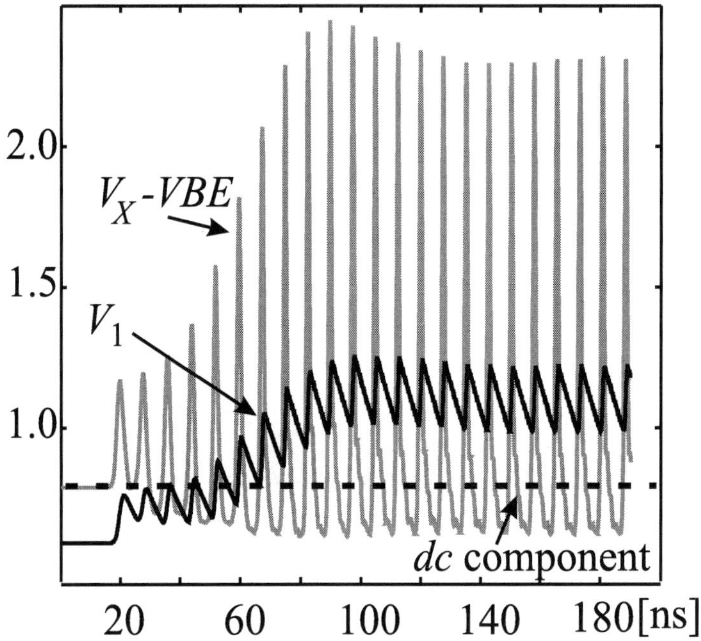

FIGURE 9.15
Simulation results of oscillator start-up. The *dc* component of $V_X - V_{BE}$ does not change; V_1 increases instead.

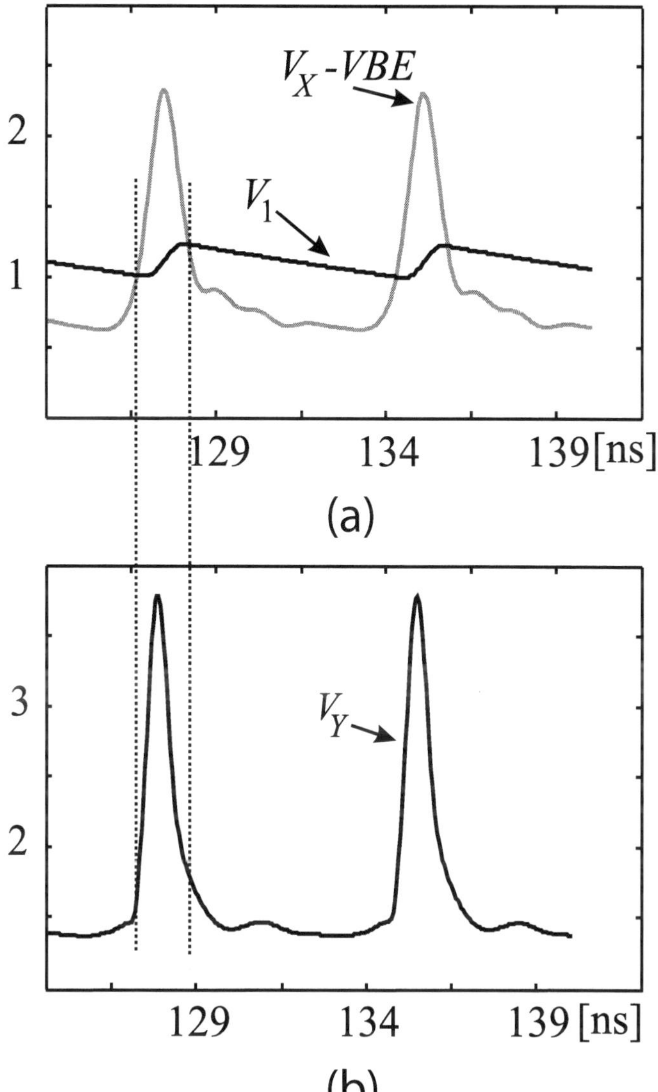

FIGURE 9.16
(a) Close-up of steady-state oscillation. The input, $V_X - V_{BE}$, is biased across V_1. (b) The output pulse is generated from the portion of the input which is above V_1. This results in the sharpening of the input pulse at the output of the amplifier. (Note: the time scale of (b) has been shifted to align with (a). The misalignment in time is due to the delay through the amplifier.)

9.2.2 Experimental Results

The steady-state soliton oscillation from the second prototype is measured using an Agilent 54855A real-time oscilloscope and is shown in Fig. 9.17(a). A periodic soliton pulse train is clearly seen. The soliton pulse repetition rate is 130 MHz (inverse of the pulse repetition period 7.7 ns) while the soliton pulse width is 827 ps FWHM, which corresponds to frequency content well into the microwave region. The corresponding power spectral density (PSD) of the steady-state signal measured using the PSD measurement capability of the oscilloscope is shown in Fig. 9.17(b). The fundamental frequency of 130 MHz corresponds to the soliton repetition rate as mentioned earlier. The significant harmonic content above 130 MHz is what determines the narrow soliton shape.

In addition to stable operation in steady-state, the microwave prototype exhibited similar spatial dynamics as the CMOS prototype as shown in Fig. 9.18. The pulse is initially not a soliton and hence sharpens into a soliton as it propagates. Once formed, the soliton continues to propagate, but experiences soliton damping due to the loss in the NLTL. The sharpening by the saturable absorption is very pronounced in this prototype, due to the sharp transition between the gain and attenuation regions of the transfer characteristics. A shorter pulse width could have been obtained, by designing the NLTL to sharpen the pulse further. Due to parasitics associated with building a discrete board, however, this was not done.

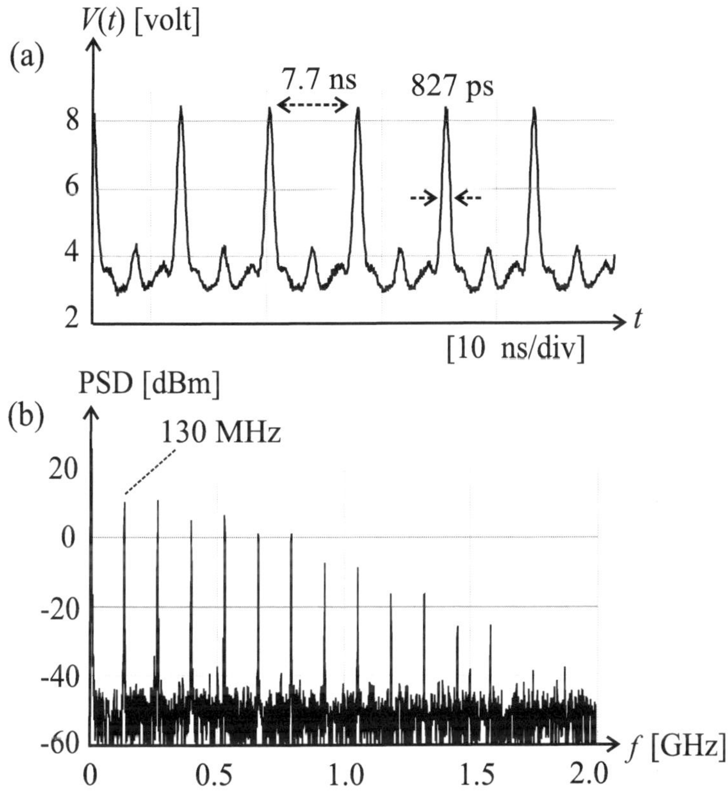

FIGURE 9.17
Steady-state soliton oscillation from the microwave soliton oscillator prototype. (a) Time-domain (oscilloscope) measurement. (b) Power spectral density obtained from the oscilloscope's PSD measurement capability. (From D. S. Ricketts, X. Li and D. Ham, "Electrical soliton oscillator," *IEEE Trans. on Microwave Theory and Tech.*, vol. 54, no. 1, pp. 373-382, Jan. 2006 © IEEE 2006.)

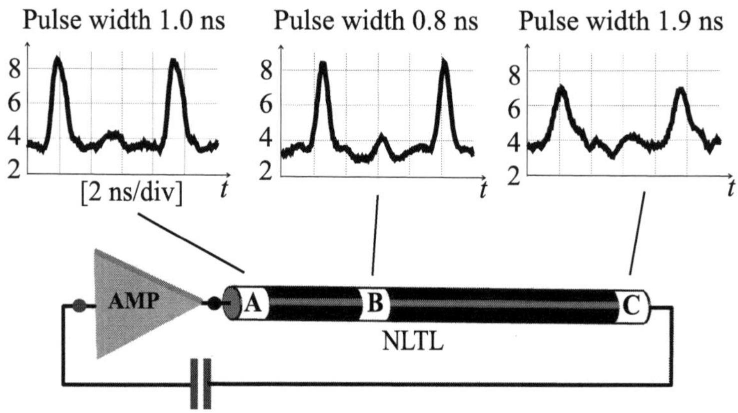

FIGURE 9.18
Spatial dynamics of the microwave prototype.

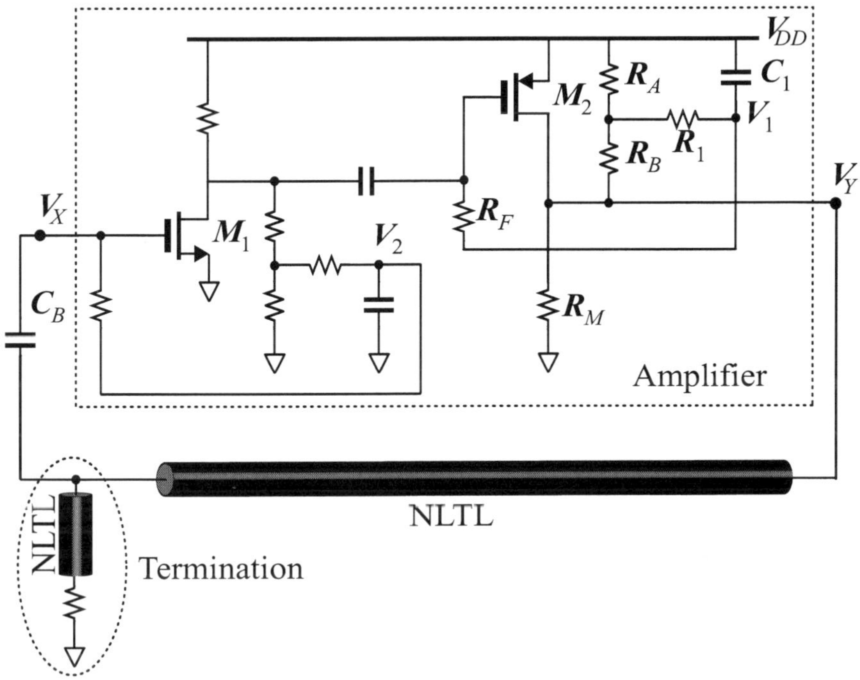

FIGURE 9.19
Detailed oscillator schematic of the chip-scale, GHz soliton oscillator. (From D. S. Ricketts, et al., "On the self-generation of electrical soliton pulses," *IEEE J. Solid State Circuits*, vol. 42, no. 8, pp. 1657-1668, Aug 2007 © IEEE 2007.)

9.3 CMOS, Chip-scale, GHz Prototype

The chip-scale soliton oscillator follows the same design procedure as the low MHz CMOS prototype. It was fabricated using an 7WL 0.18 μm analog RF CMOS process. Figure 9.19 redraws the detailed schematic for the oscillator.

9.3.1 Oscillator Implementation

The challenge in constructing the chip-scale prototype, in addition to the stabilization of solitons, was constructing an amplifier and NLTL which would support shorter soliton widths (300 ps) and higher repetitions rates (GHz). The technology chosen provided the bandwidth necessary to amplify pulses down to approximately 100 ps. The main challenge then was in the design of the NLTL.

The target pulse width for the chip-scale soliton oscillator was a few hundred picoseconds. This rather large pulse width was needed to ensure that the spectral content of the oscillating pulses fell below the bandwidth limitations of our measurement system (7 GHz). This bandwidth limitation was imposed by the choice of a real-time oscilloscope; over-sampling oscilloscopes could have easily measured signals up to 50 GHz. The reason for choosing a real-time oscilloscope was the desire to capture time varying events that might occur in an unstable soliton oscillation. An oversampling scope would have not been able to accurately capture these events.

With the target pulse width chosen, (9.1) and (9.2) were used to determine the inductance and nominal capacitance of the *pn* junction varactor (the parameter b is dictated by the technology). The first design constraint was to minimize the inductance for on-chip implementation. The constraining parameters are the target pulse width, the repetition rate and the nominal impedance of the NLTL. The latter is important, since the amplifier cannot drive an arbitrarily low impedance. In this chip-scale prototype an impedance on the order of 100 Ohms was chosen (in the discrete prototypes, large inductor values were practical, allowing characteristic impedances of several hundred ohms). The final values were a nominal capacitance of a few hundred femtofarads and an inductance of 1–3 nH.

These values of inductance are easily integrated on chip with spiral inductors. Unfortunately, the typical spiral inductor Q of less than 10 ($Q = \frac{\omega L}{R}$) proved to be too lossy to sustain oscillations. In simulation, we found that a Q of at least 18 was needed to produce large-signal soliton oscillations. Several options were available, such as plating standard spiral inductors, using two thick-metal inductors in parallel, and also using bondwire inductors. The first two options were not available or easily compatible with the technology. The latter, bondwire inductors, are easy to implement using a standard bondwire machine. Studies on bondwire inductors have shown Qs greater than 20 [66], making them ideal for our NLTL. The only downside is their low inductance, approximately 1 nH per mm of length. Given our target inductance of 1–3 nH, this was not a problem.

The final implementation of the chip-scale soliton oscillator is shown in Fig. 9.20. Figure 9.20(a) shows a micrograph of the implemented chip-scale soliton oscillator, which consists of an IC mounted on an in-house fabricated glass substrate [38]. The NLTL inductors were created by bonding gold wires back and forth between the pads on the IC and metallic pads on the glass substrate. Bonding was done by using an automated bonding machine to ensure consistency in wire length. It was estimated that the variation from inductor to inductor was less than 5%. Magnetic couplings between adjacent bonding wires can be estimated via electromagnetic simulations to choose proper lengths of the bonding wires to generate a desired inductance. Figure 9.20(a) shows a side view of the array of bondwire inductors.

The utilization of the bond wires as NLTL inductors allowed us to test various different NLTL characteristics (inductance value and NLTL length) in the soliton oscillator via post-fabrication adjustment. At the same time, however, the pads on the IC necessary for bonding add parasitic linear capacitors and hence reduce the nonlinearity of the NLTL (resulting in a reduced pulse sharpening by the NLTL).

9.3.2 Test and Measurement

The testing of the chip-scale prototype was done using the test board shown in Fig. 9.21(a). The glass substrate was mounted in the test package and then the die glued onto the glass substrate. This provided a firm fixture for the automatic wirebonding. The *dc* biasing lines were connected to the pins of the test package and routed to SMA connectors on the test board. Decoupling capacitors were placed on the board to reduce noise coupling. Figure 9.21(b) shows how the chip was measured. A GGB PicoProbe with a bandwidth of 26 GHz (Model 35) was used to probe the pads of the glass substrate directly. This allowed for measurement of the oscillation without the need for any additional buffers. An Agilent 54855A real-time oscilloscope was used to capture the signal from the PicoProbe.

9.3.3 Experimental Results

The chip-scale system of Fig. 9.20 indeed self-starts from noise, leading to a self-sustained stable soliton oscillation in steady state. Figure 9.22 shows the soliton train produced by a sample, measured at one point on its NLTL. The soliton train has a pulse repetition rate of 1.14 GHz and a pulse width of 293 ps at full width half maximum (FWHM). Several different samples obtained by varying the NLTL characteristics (different lengths and inductance values) via bond wire adjustment led to pulse repetition rate ranging from 0.9 to 1.9 GHz and pulse width ranging from 293 ps to 400 ps.

To examine how the pulse shape changes as it circulates in the oscillator loop in steady state oscillation, the voltage signals were measured

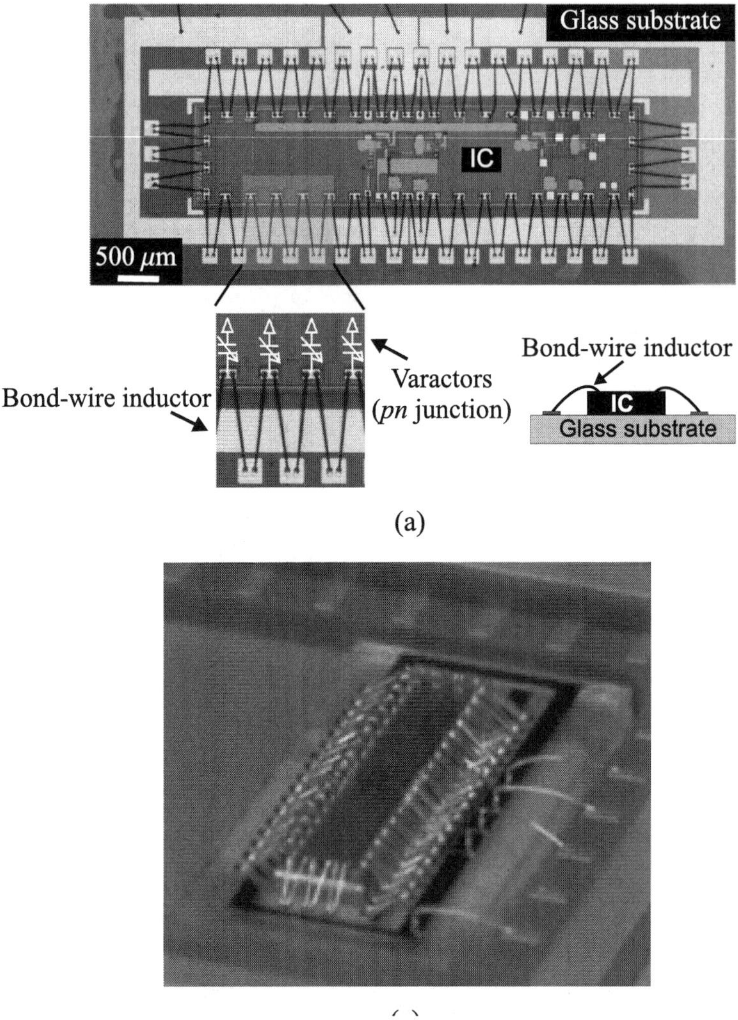

FIGURE 9.20
Chip-scale soliton oscillator implementation. (a) Micrograph of the chip-scale soliton oscillator, which consists of a CMOS IC mounted on a glass substrate. NLTL inductors are created by bonding gold wires back and forth between the IC and substrate. The IC contains the amplifier and varactors. (b) Array of bondwire inductors. (From D. S. Ricketts, et al., "On the self-generation of electrical soliton pulses," *IEEE J. Solid State Circuits*, vol. 42, no. 8, pp. 1657-1668, Aug 2007 © IEEE 2007.)

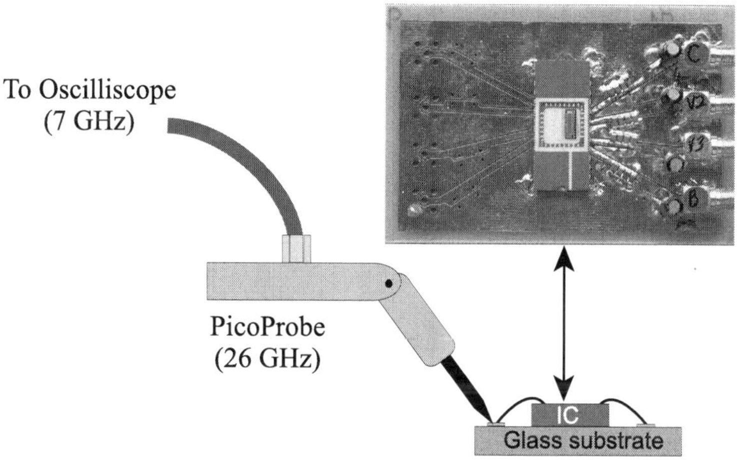

FIGURE 9.21
Test setup consisting of *dc* testing board and a 26 GHz GGB PicoProbe.

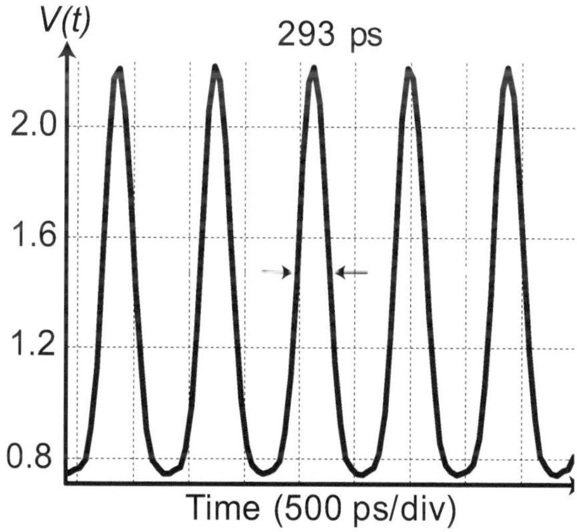

FIGURE 9.22
Steady-state soliton oscillation. The pulse repetition rate and width are 1.14 GHz and 293 ps (FWHM), respectively. (From D. S. Ricketts, et al., "On the self-generation of electrical soliton pulses," *IEEE J. Solid State Circuits*, vol. 42, no. 8, pp. 1657-1668, Aug 2007 © IEEE 2007.)

at various points on the NLTL using the real-time oscilloscope. Figure 9.23 shows the steady-state voltage signals measured at three different NLTL positions. The pulse at the immediate output of the amplifier, V_Y, which is 429 ps wide (FWHM), is not an exact soliton, and hence, sharpens into a soliton as it propagates down the NLTL. Through this pulse compression process, the soliton is formed at the eighth section achieving the minimum pulse width of 385 ps. Once the soliton is formed, as it propagates farther down the NLTL, it no longer sharpens and the loss on the NLTL becomes the dominant process; the soliton undergoes soliton damping [42] where its amplitude and speed decrease and its width increases while the spatial confinement of wave energy is still maintained in the soliton pulse. At V_X the pulse width has been reduced to 498 ps. The pulse is then sharpened by the saturable absorption mechanism in the amplifier back to a width of 429 ps at V_Y. This spatial process mirrors the dynamics of the low MHz prototype in Subsec. 9.1.2.5.

9.4 Summary

The prototypes presented in this chapter have not only demonstrated the robust self-generation of electrical solitons, but have also shown many of the fascinating dynamics of solitons, such as amplitude-dependent velocity, pulse sharpening, and soliton collisions. In addition, the underlying stabilizing mechanisms have been clearly and explicitly demonstrated, validating the concepts developed in Chapter 8.

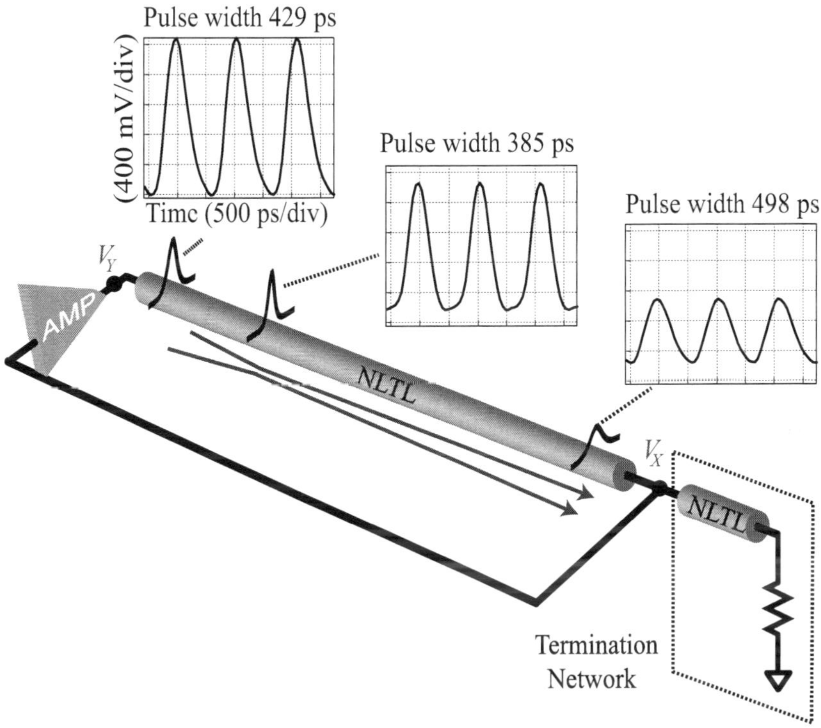

FIGURE 9.23
Spatial dynamics of the chip-scale soliton oscillator in steady state. The pulse at V_Y is 429 ps wide (FWHM). As the pulse propagates down the NLTL, it is compressed into a soliton at the eighth section of the NLTL, reaching a minimum width of 385 ps. Once formed into a soliton, it propagates as a soliton and experiences unique soliton damping due to the loss in the NLTL further propagating down the NLTL. At V_X the pulse has widened to a width of 498 ps width. (From D. S. Ricketts, et al., "On the self-generation of electrical soliton pulses," *IEEE J. Solid State Circuits*, vol. 42, no. 8, pp. 1657-1668, Aug 2007 © IEEE 2007.)

10

The Reflection Soliton Oscillator

O. Ozgur Yildirim
Harvard University

CONTENTS

This chapter presents another soliton oscillator architecture called the reflection soliton oscillator. In the circular soliton oscillator, which is described in the previous chapter, a soliton pulse circulates on a circular loop, consisting of a nonlinear transmission line (NLTL) and a two-port amplifier with level dependent gain function [see Fig. 10.1(a)]. In the reflection soliton oscillator, instead of circling, a soliton pulse travels back and forth on an NLTL. The circuit consists of a one-port reflection amplifier and an NLTL as shown in Fig. 10.1(b). One side of the NLTL is connected to the reflection amplifier, which provides gain to the soliton to compensate the transmission line loss and ensures oscillation stability by suppressing undesired perturbations via the level-dependent gain function in the same manner as in the circular soliton oscillator. The other end of the NLTL is an electrical open, where the soliton reflects back. This chapter presents the design, fabrication, and experimental results of the reflection soliton oscillator.

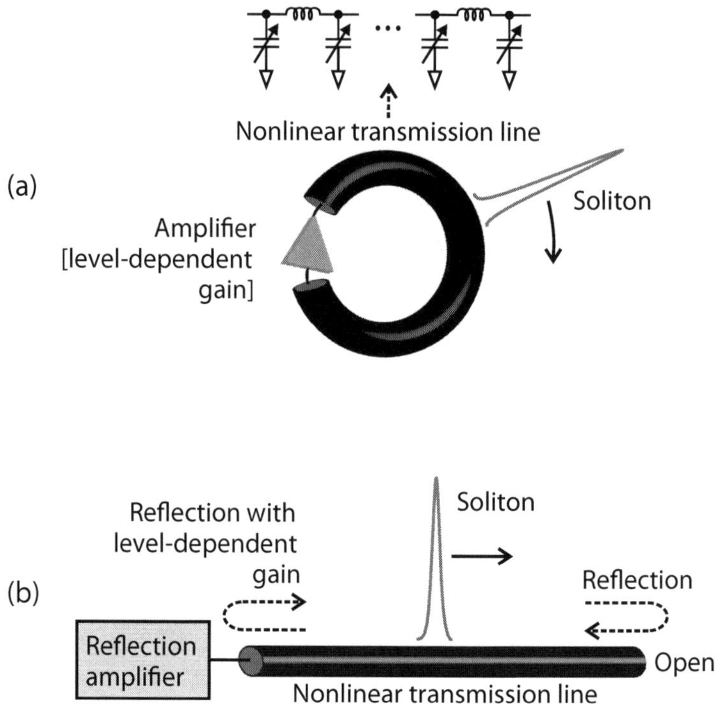

FIGURE 10.1
Block diagram of: (a) the circular soliton oscillator [16]; (b) the reflection soliton oscillator [67].

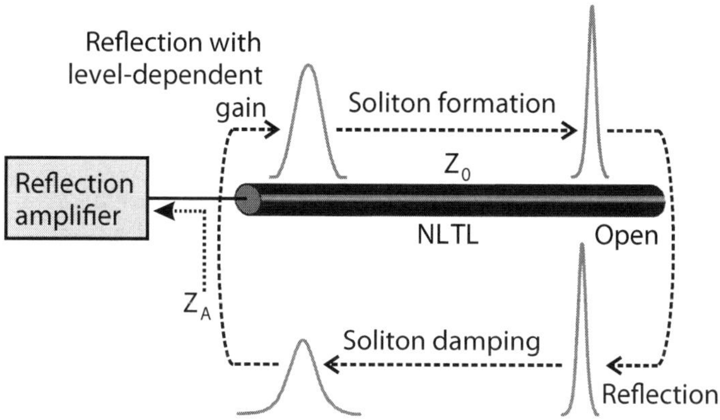

FIGURE 10.2
Steady-state dynamics in the reflection soliton oscillator.

10.1 Operating Principle

The reflection soliton oscillator consists of an NLTL terminated with a one-port amplifier at one end and an electrical open at the other end [see Fig. 10.1(b), or Fig. 10.2]. In the steady state (oscillation startup will be addressed in Sec. 10.2), a pulse travels back and forth on the NLTL as shown in Fig. 10.2. The pulse reflected at the amplifier end travels down the NLTL towards the open, during which the NLTL compresses the pulse, forming it into a sharp soliton. As the soliton reaches the open, it is reflected to travel back towards the amplifier, during which the soliton lowers amplitude and broadens width in a particular manner [42] due to loss in the NLTL. Once the damped soliton reaches the amplifier, a reflection with level-dependent gain occurs, which compensates loss and reinforces oscillation stability. The reflected pulse is not a soliton, for the amplifier deforms the soliton shape. The reflected, reenergized pulse repeats the spatial dynamics described above, again forming into a soliton as it propagates down the NLTL towards the open.

10.1.1 Reflection at the Amplifier End

The reflection with level-dependent gain at the amplifier end is achieved as follows. The output impedance of the amplifier, Z_A, seen by the NLTL, is a function of voltage V_A at that output node. If the average characteristic impedance[1] of the NLTL is denoted as Z_0, the reflection coefficient, Γ, at the amplifier end, with a voltage of V_A, is given by

$$\Gamma(V_A) = \frac{Z_A(V_A) - Z_0}{Z_A(V_A) + Z_0} \tag{10.1}$$

Since $Z_0 > 0$, we see that

$$|\Gamma(V_A)| \begin{cases} > 1 & \text{(gain)} & \text{if } Re\{Z_A(V_A)\} < 0 \\ < 1 & \text{(attenuation)} & \text{if } Re\{Z_A(V_A)\} > 0 \end{cases} \tag{10.2}$$

Therefore, for the amplifier to provide a gain for a voltage larger than a certain threshold and to attenuate a signal smaller than the threshold, we are to have $Re\{Z_A\} < 0$ for V_A above the threshold, and $Re\{Z_A\} > 0$ for V_A below the threshold, as hypothetically drawn in Fig. 10.3. This is how the reflection with level-dependent gain is attained. The design of such an amplifier will be detailed in Section 10.2.

Let us examine what would happen if the level-dependent gain is not in force, or more specifically, if $Re\{Z_A\} < 0$ regardless of V_A, and all signal levels receive gain upon reflection from the amplifier end. Suppose a desired soliton pulse and a small parasitic soliton pulse produced by a perturbation (e.g., noise) in the oscillator [see ①, Fig. 10.4(a)]. Both of these pulses will be reflected with gain at the amplifier end, and the parasitic pulse will persist, as the amplifier in this scenario does not attenuate low-level signals [see ②, Fig. 10.4(a)]. Since the two reflected pulses have different amplitudes, they propagate at different velocities (a taller pulse propagates faster than a shorter pulse — this is a key soliton property), thus eventually colliding with each other. This collision, which is a nonlinear process, causes pulse amplitude modulation at the moment of collision [see ③, Fig. 10.4(a)], and pulse position modulation after the collision [see ④, Fig. 10.4(a) — the taller pulse would appear as the dashed one in the different position, were it not for the collision], which is another key soliton property. Therefore, the pulse train produced with gain at all signal levels exhibits significant variations in pulse amplitude and repetition rate. See Fig. 10.4(b) for an unstable

[1]The NLTL's characteristic impedance at a given point varies with the voltage applied to the varactor at that point, due to the voltage dependence of the varactor capacitance. Here we take the average characteristic impedance seen by a signal to capture the essence in a simple, albeit not the most accurate, manner.

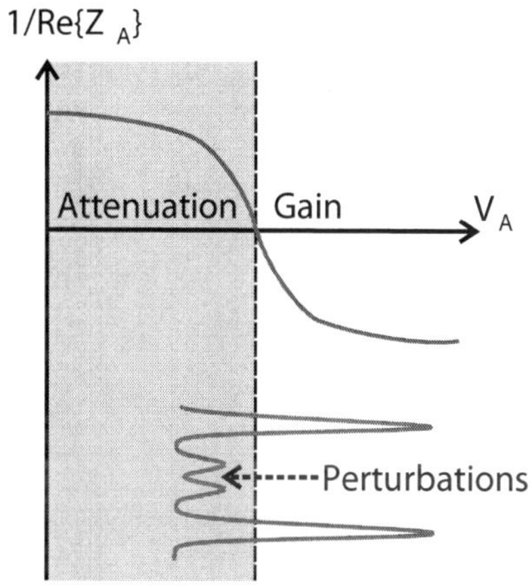

FIGURE 10.3
Hypothetical depiction of the desired Z_A as a function of V_A, which yields level-dependent gain upon reflection of a pulse at the amplifier end. This characteristic is needed for soliton oscillation stability.

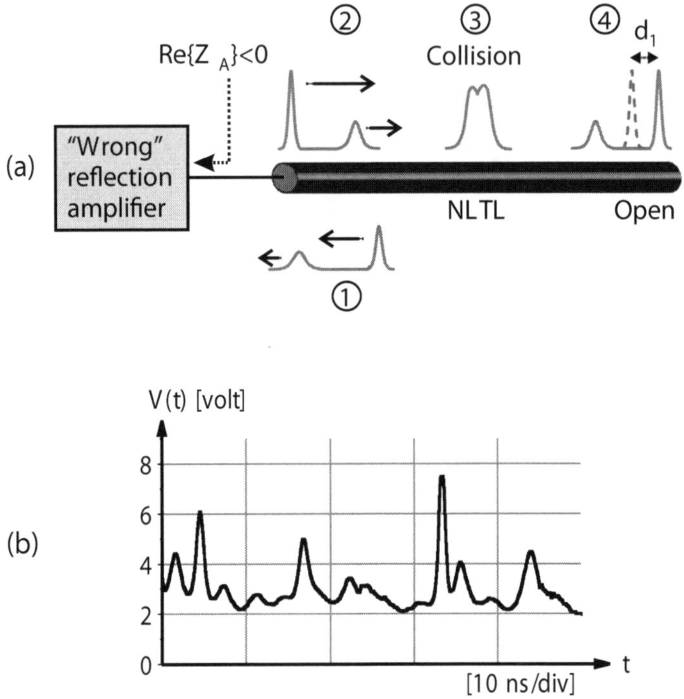

FIGURE 10.4
(a) Buildup of soliton oscillation instability when gain is provided for all signal levels. (b) Measured unstable oscillation when gain is provided for all signal levels.

oscillation observed in one of actual prototypes. In contrast, when level-dependent gain is in force, small perturbation falls in the attenuation region to be suppressed (see Fig. 10.3), no soliton collision occurs, and the instability is prevented.

10.1.2 Reflection at the Open End

The pulse reflection at the open end of the NLTL, illustrated in Fig. 10.5, is interesting and useful, for it offers an extra pulse sharpening mechanism, in addition to the sharpening that occurs during pulse propagation on the NLTL.

At the open end, the incoming pulse is reflected into a same shape pulse. The incoming and reflected pulses (voltages) superpose to increase the joint voltage at the open end. If a linear line is used, the joint voltage at the open end is twice the incoming voltage. In case of the NLTL, the joint voltage is *larger* than twice the incoming voltage. This is due to the voltage dependence of the capacitance of the varactor at the open end (see inset of Fig. 10.5 for a depiction of the varactor's capacitance-voltage characteristic). The superposed (thus increased) voltage reduces the capacitance, and to conserve the total energy despite the reduced capacitance, the superposed voltage should be larger than twice the incoming voltage [10]. In the measurement (see Section 10.3), the joint pulse amplitude is 2.4 times larger than the incoming pulse amplitude.

This pulse amplification at the open end by a factor larger than two translates to pulse sharpening. In the NLTL consisting of inductor-varactor sections (see inset of Fig. 10.5), if loss is ignored (this is justifiable, as the qualify factor of the NLTL used is around 100 in the operation frequency range), the following holds for any n, according to KVL:

$$L\frac{\partial I_{n+1}(t)}{\partial t} = V_n(t) - V_{n+1}(t) \tag{10.3}$$

The time integration of the left-hand side from $t = -\infty$ to ∞ reduces to zero, as pulses die down at $t = \infty$ and do not exist $t = -\infty$. Therefore, the time integration of the right-hand side from $t = -\infty$ to ∞ is zero, or,

$$\int_{-\infty}^{\infty} V_n(t)dt = \int_{-\infty}^{\infty} V_{n+1}(t)dt \quad \text{(for any } n\text{)} \tag{10.4}$$

Therefore, the time integration of an arbitrary pulse voltage $V_n(t)$ at the n-th node is constant, independent of the position n [10]:

$$\int_{-\infty}^{\infty} V_n(t)dt = \text{constant} \tag{10.5}$$

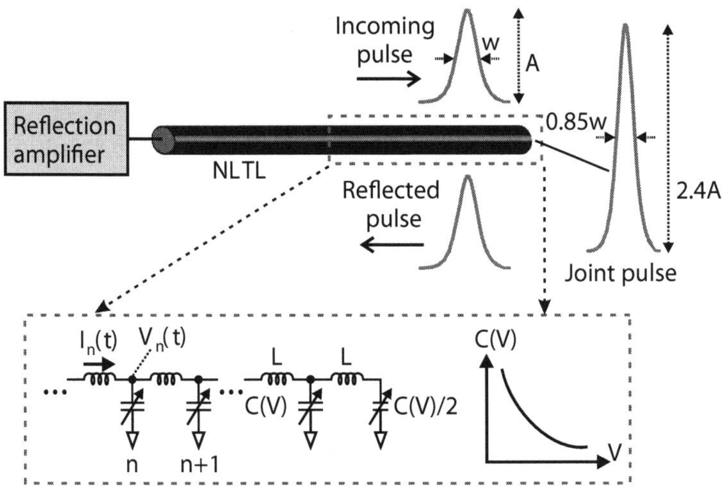

FIGURE 10.5
Reflection of the soliton at the open end of the NLTL. The joint pulse at the open end is sharper and more than twice taller than the incoming soliton.

In other words, if one measures the area under the time-domain voltage waveform $V_n(t)$, the area will remain the same, independently of n. Therefore, the time-domain area of the joint voltage pulse at the open end should be twice the time-domain area of the incoming voltage pulse, whose amplitude is A and temporal width is W. Since the time-domain area of a pulse may be approximated as a product of its amplitude and temporal width, the time-domain area of the joint voltage pulse is $2AW$. Therefore, the amplitude $2.4A$ of the joint pulse corresponds to the temporal width of $0.83W$ (in the experiment of Section 10.3 the measured width is $0.85W$). Thus, a pulse sharpening occurs at the open end of the NLTL, as the joint pulsewidth $0.83W$ at the open end is smaller than the incoming pulse-width W.

10.2 Amplifier Design

10.2.1 Need for an Adaptive Bias Scheme

As discussed in the previous section, the reflection amplifier in Fig. 10.2 should be capable of level-dependent amplification such that small perturbations are rejected to ensure oscillation stability while the main portion of the soliton is amplified to compensate loss. An important notion is that this level-dependent gain is what is required in the *steady state.* In contrast, in the *initial startup transient*, the level-dependent gain should be actually avoided, for small perturbations should be amplified to enable oscillation startup.

To achieve these two contradicting gain characteristics in the same amplifier, an adaptive bias scheme can be employed to shift the gain characteristic of the amplifier from full gain to level-dependent gain as the oscillation grows from an initial transient into a steady-state pulse train. Such an adaptive bias scheme has been actually implemented in various forms in all of the previous pulse oscillators discussed earlier in the book [16, 22, 64]. In this work, an adaptive bias scheme suitable for the reflection soliton oscillator is also devised, which will be explained now.

10.2.2 Reflection Amplifier with an Adaptive Bias

Figure 10.6 shows the one-port reflection amplifier, which achieves, through an adaptive bias scheme, the full gain during the initial startup transient and the level-dependent gain during the steady state. The ba-

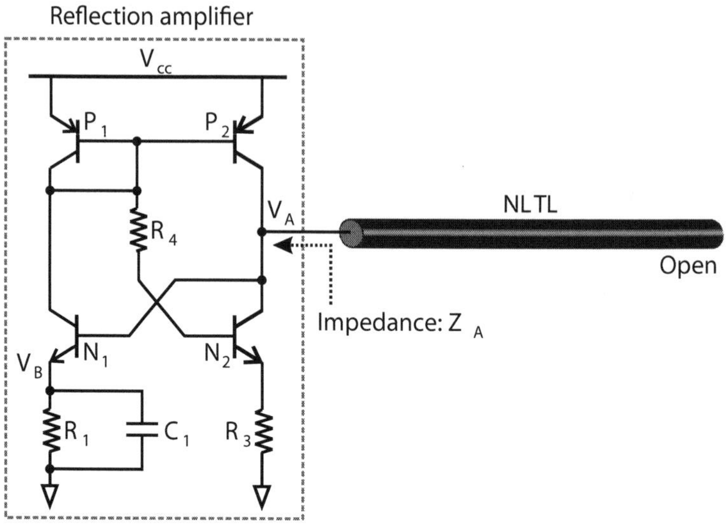

FIGURE 10.6
The detailed schematic of one-port reflection amplifier that executes the level-dependent gain function.

sic circuit arrangement is two inverters put back to back: transistors $\mathbf{N}_1$ and $\mathbf{P}_1$ form one inverter; transistors $\mathbf{N}_2$ and $\mathbf{P}_2$ form the other inverter. The underlying design idea is that the real part of the output impedance of the amplifier, $Re\{Z_A\}$, seen by the NLTL, can assume positive or negative values, depending on the ON/OFF status of $\mathbf{N}_1$, that is, depending on the voltage difference between V_A and V_B. If $\mathbf{N}_1$ is ON with $V_A > V_B$, the back-to-back inverters are in force and a positive feedback is set up, yielding $Re\{Z_A\} < 0$. If $\mathbf{N}_1$ is OFF with $V_A < V_B$, the back-to-back inverters do not work, leading to $Re\{Z_A\} > 0$.

In the beginning of the startup transient, V_A and V_B assume certain initial bias values. The amplifier is arranged in such a way that in this initial stage $V_A > V_B$ and $\mathbf{N}_1$ is ON, thus ensuring $Re\{Z_A\} < 0$ and enabling an oscillation startup. Fig. 10.7(a) shows the real part of the simulated small signal impedance Z_A at various *dc* voltages V_A applied at the output of the amplifier. As can be seen, the initial V_A bias sits in the gain region where $Re\{Z_A\} < 0$, and thus small perturbations around the bias receive gain to grow into a pulse train.

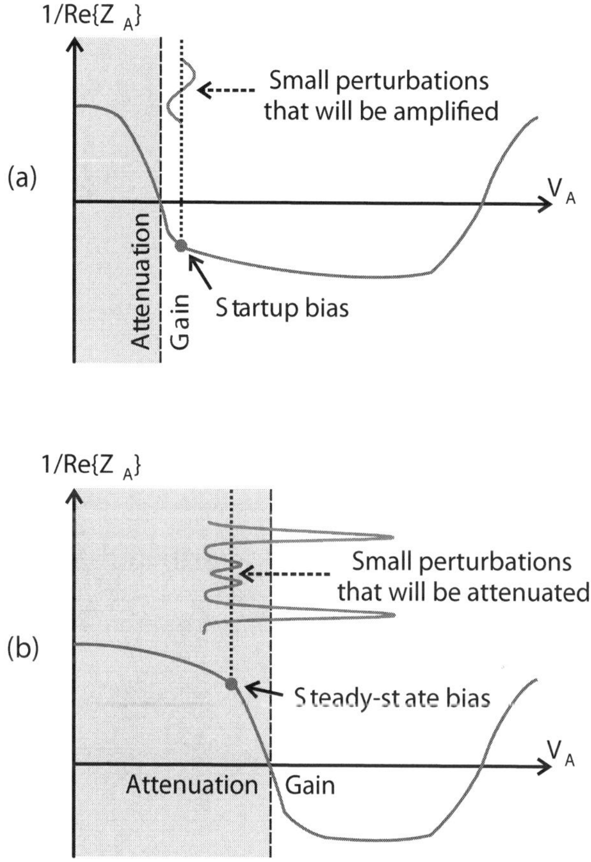

FIGURE 10.7
$1/Re\{Z_A\}$ versus V_A: (a) Initial startup stage; (b) The steady state.

As the oscillation grows, a train of pulses starts appearing at the output node of the amplifier. See Fig. 10.8 in conjunction with Fig. 10.6 for the following description. When a pulse arrives at the amplifier from the NLTL, V_A goes up. Following V_A, the emitter voltage V_B of $\mathbf{N}_1$ is increased, charging capacitor C_1. Due to $\mathbf{N}_1$, the time constant associated with this charging of C_1 is much smaller than R_1C_1, the time constant of the R_1-C_1 network.[2] Now as the pulse leaves the amplifier and V_A is decreased, $\mathbf{N}_1$ is turned off because V_B is still high, and the charge stored on C_1 starts being discharged through the R_1-C_1 network with a time constant of R_1C_1: the discharging of C_1 in the absence of a pulse is slower than its charging in the presence of a pulse. The value of R_1C_1 is set larger than the pulse repetition period, so that by the time the next pulse arrives, the discharging has not been sufficient to empty what was charged during the previous pulse event (see ① and ②, Fig. 10.8), and thus, V_B has been increased overall (see Fig. 10.8). As this process is repeated with continued arrivals and departures of pulses, the overall charges in C_1 are increased at every pulse event, and the *dc* average (bias) of V_B goes up. In this way, once the steady state is reached (it will be explain shortly how the *dc* average of V_B eventually settles to a constant value), the bias of V_B assumes a value larger than that during the initial startup.

Due to this increased V_B, in the steady state, turning on $\mathbf{N}_1$ to generate a negative $Re\{Z_A\}$ and to produce gain requires a higher level of voltage at V_A (see Fig. 10.8). This corresponds to the fact that in the steady state, the $1/Re\{Z_A\}$ versus V_A curve is shifted to the right as hypothetically shown in Fig. 10.7(b): negative $Re\{Z_A\}$, or gain, is attainable at higher voltages. The steady-state bias (*dc* average) of V_A has been shifted to the right as well due to the increased pulse height, but this shift is not as large as the shift in the impedance curve, and thus, the steady state bias lies in the attenuation region. Overall, the steady state pulses lie across the attenuation and gain regions as shown in Fig. 10.7(b) or in Fig. 10.8, and only the higher portion of the pulses will be amplified while their lower portion and any small undesired perturbations will be suppressed: the level-dependent gain has thus been set up in the steady state.

Shortly before, it is asserted that the increase in the *dc* average of V_B during the initial startup transient eventually settles down to a con-

[2]In the small signal analysis, the time constant associated with the C_1 charging can be approximated as $R_1C_1/(1+g_mR_1)$, where g_m is the transconductance of $\mathbf{N}_1$. The charging time constant is smaller than R_1C_1 by a factor of $1+g_mR_1$. In our large signal scenario, although the equation above is not exactly the charging time constant, it is a sufficient indication that the charging time constant is smaller than R_1C_1.

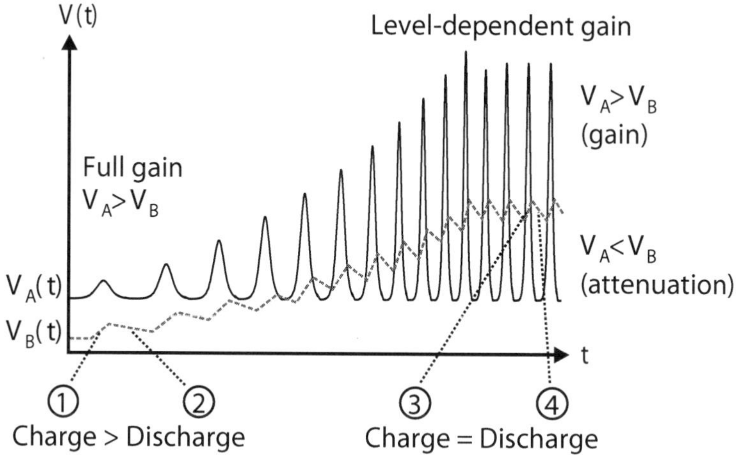

FIGURE 10.8
Illustration of a startup transient into steady state.

stant value, which can be explained as follows. During the very initial transient, the C_1 charging is faster than the C_1 discharging due to the strong action of $\mathbf{N}_1$ that is ON during the pulse event, thus causing the increase in V_B. As the steady state is approached, $\mathbf{N}_1$ turns on during the higher portion of a pulse, but turns off during the lower portion of the pulse, and thus, the effect of $\mathbf{N}_1$, which is not always ON even during the pulse event, has become weaker, and thus, the C_1 charging is not as fast. Due to this slowing down in the C_1 charging rate, at a certain point, the charging during the pulse event and the discharging during the absence of the pulse balance each other out (see ③ and ④, Fig. 10.8), thus causing the *dc* average of V_B to settle at a constant value.[3]

From the foregoing description, it is clear that R_1C_1 should be larger than the pulse repetition period T_0 to ensure a slow discharging between two adjacent pulse events, so that V_B can grow during the initial startup. In this circuit, $R_1C_1 \approx 1.7T_0$. Note, however, that too large a value of R_1C_1 should be avoided, for it causes Q-switching, or relaxation oscillation [64].

[3]The charging rate is still larger than the discharging rate, but the charging time is smaller than the discharging time, due to the small duty cycle of the pulse train.

10.2.3 Improved *R-C* Network

In the initial transient, it is common to see two competing large pulses that appear successively well within the intended pulse repetition period. The amplifier should be able to select only one of them, while eliminating the other, so that the oscillator can eventually produce a periodic train of soliton pulses with a constant amplitude and a constant pulse repetition rate. Consider two such pulses. If the increase in V_B during the first pulse is not fast enough, V_B is still low for the second pulse, and thus, the second pulse receives a large enough gain to be undesirably sustained. To prevent this, V_B should rise fast enough during the first pulse so that by the time the second pulse arrives, V_B is high enough to introduce enough loss for the second pulse to eliminate it. To this end, an R_2-C_2 network is introduced in series with the original R_1-C_1 network, as shown in Fig. 10.9, with $R_2C_2 \approx 0.06R_1C_1$. This extra *R-C* network allows for V_B to track V_A fast for the first pulse, so that the second pulse can be suppressed.

10.3 Experiments

For proof of concept, a reflection soliton oscillator prototype in a discrete form was constructed (see Fig. 10.10), whose output signal lies across a lower end of the microwave spectral region. The one-port reflection amplifier was implemented using bipolar transistors[4] with maximum unity current gain frequencies of 5 GHz. The NLTL was constructed by forming an *LC* ladder network with 24 inductor-varactor sections.[5] A 6-V *dc* power supply was used.

The prototype oscillator self-starts from ambient noise, and always reaches a steady state where a stable, periodic train of pulses is self-sustained. The oscillation waveforms are measured using an Agilent Infiniium 54855A 6-GHz oscilloscope in conjunction with Agilent 1156A active probes with high impedance (100 kΩ).

Figure 10.11 shows a start-up transient measured at the open end of the NLTL. As discussed earlier, this signal corresponds to the nonlinear superposition of the incoming and reflected solitons. As the pulse amplitude grows with time, its width is decreased. This amplitude-dependent

[4] For *npn* transistors, Philips BFT25A bipolar transistors were used. For *pnp* transistors, Philips BFT93 bipolar transistors were used.

[5] For inductors, API Delevan 4426 series air core inductors were used. For varactors, Zetex ZMV930 varactors were used.

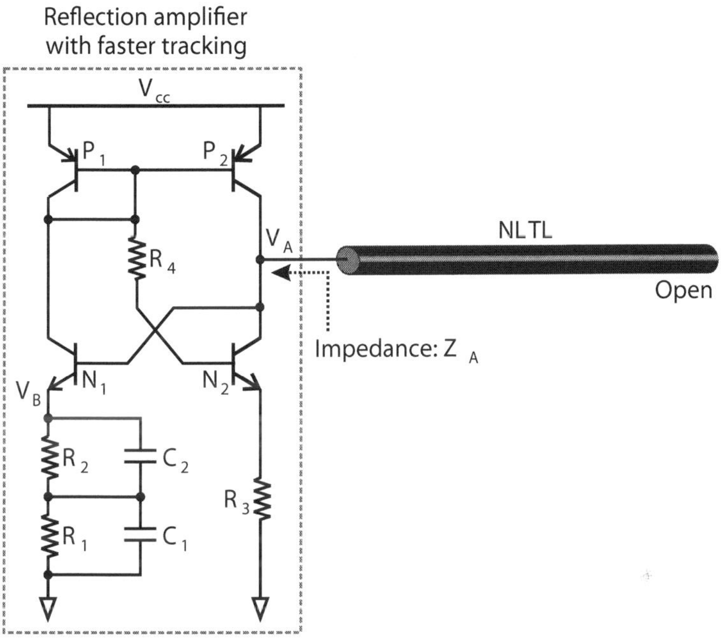

FIGURE 10.9
The reflection amplifier with an improved *R-C* network, which provides faster tracking.

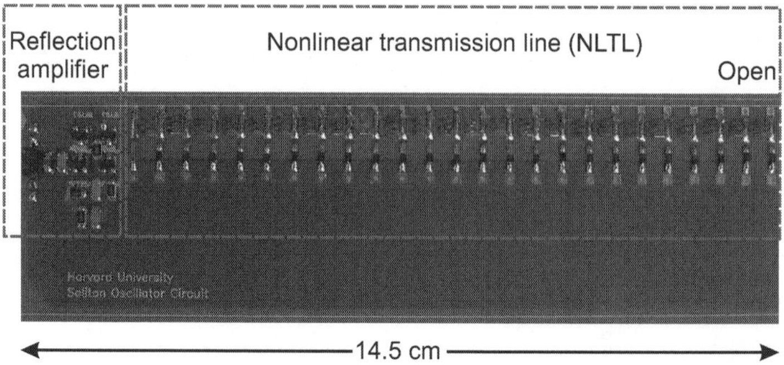

FIGURE 10.10
Reflection soliton oscillator discrete prototype.

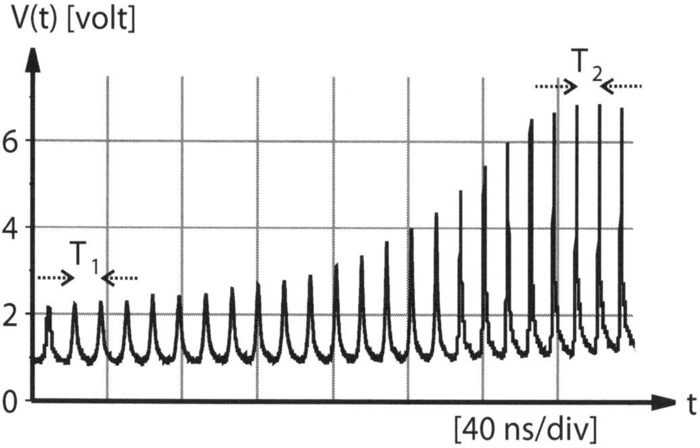

FIGURE 10.11
Measured initial startup transient at the open end of the NLTL.

width is a key soliton property. It is also observed that with the growing pulse amplitude, the pulse repetition rate is increased: e.g., in the figure, $T_1 > T_2$. This is due to another soliton property that a taller soliton propagates faster.

Figure 10.12(a) shows a stable, periodic train of pulses measured at the open end of the NLTL in the steady state, and Fig. 10.12(b) shows its Fourier transform into frequency domain, where the Fourier transform is performed using the FFT capability of Agilent Infiniium 54855A oscilloscope. The pulsewidth and pulse repetition period are 445 ps and 9.7 ns (4.6% of duty cycle; 103 MHz of pulse repetition rate; many higher harmonics well into the GHz frequencies).

Figure 10.13 shows measured $V_A(t)$ and $V_B(t)$ from the initial startup to the steady state (V_A, V_B, transistors, and other circuit components mentioned here are with reference to Fig. 10.9), which is the experimental counterpart of the hypothetical illustration of the same dynamics in Fig. 10.8. This measurement confirms the machinery of the reflection amplifier described in Section 10.2. First, initially $V_A(t)$ is larger than $V_B(t)$, and thus, $\mathbf{N}_1$ is ON and $Re\{Z_A\} < 0$, enabling the oscillation startup. Second, as the pulse train is formed and grows, the capacitors in the RC network accumulate more charges, increasing the *dc* average of $V_B(t)$. Third, in the steady state, the lower portion of $V_A(t)$ goes below $V_B(t)$, turning off $\mathbf{N}_1$, and yielding $Re\{Z_A\} > 0$. Therefore, the lower portion of $V_A(t)$ is attenuated. In contrast, a higher portion of $V_A(t)$ lies above $V_B(t)$, and thus, $\mathbf{N}_1$ is turned on, and $Re\{Z_A\} < 0$. Therefore, the higher portion of V_A receives gain. The level-dependent gain is thus properly set up during the steady state, leading to the stable pulse oscillation.

Figure 10.14 shows how the pulse shape changes in the steady state as it travels back and forth on the NLTL. At a given time, there is only one pulse traveling on the NLTL. The middle of Fig. 10.14 shows voltage $V_{16}(t)$ measured at the 16th node on the NLTL, where the node number is counted from the amplifier end. Pulse ① propagates backwards (toward the amplifier). This pulse is reflected by the amplifier, and propagates forward (toward the open end), reemerging at the 16th node as pulse ②. This pulse then is reflected at the open end, and travels backwards again, thus reappearing at the 16th node as pulse ③. At the amplifier end (see Fig. 10.14, left) and at the open end (see Fig. 10.14, right), the forward and backward waves superpose, and we see their combined effect.

As can be seen in Fig. 10.14, the pulse at the amplifier end (1.45 ns is the width of the joint pulse; the estimated width of the actual forward pulse without the superposition is 900 ps) is sharpened into a soliton, traveling down the NLTL, leading to a pulsewidth of 525 ps at the 16th

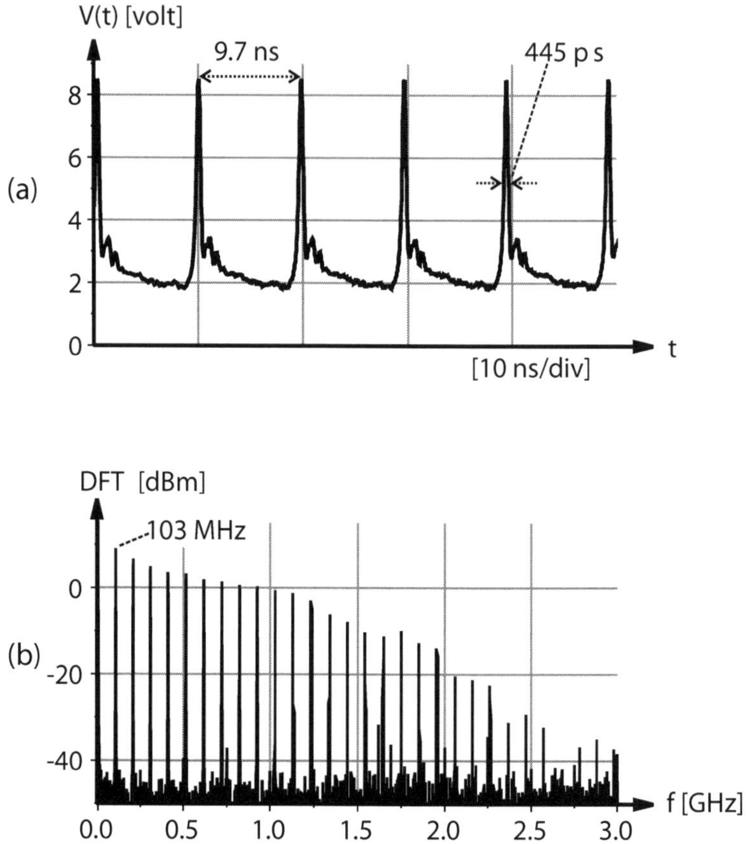

FIGURE 10.12
(a) Measured steady-state oscillation at the open end of the NLTL. (b) Fourier transform of the steady-state waveform.

node (②, ④). As this soliton reaches the open end, the nonlinear superposition occurring during the reflection (see Section 10.1.2) amplifies the pulse amplitude by 2.4 times and further sharpens the pulsewidth to 445 ps. The backward soliton has lower amplitude and broader width (see ①, ③ in comparison to ②, ④) due to soliton damping [42].

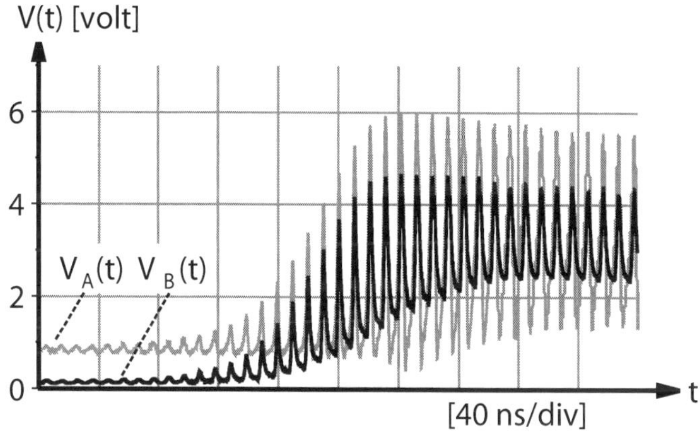

FIGURE 10.13
Measured $V_A(t)$ and $V_B(t)$ from the startup to the steady state.

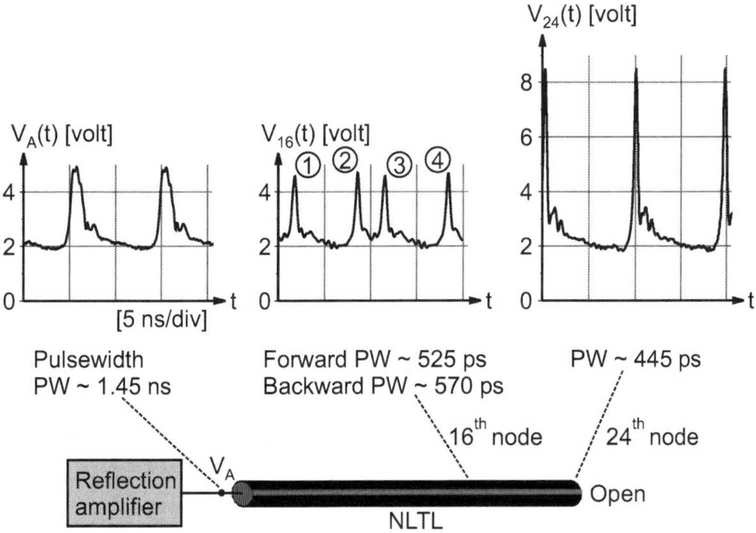

FIGURE 10.14
Measured steady-state oscillation at various points. The NLTL substantially sharpens the pulse.

10.4 Comparison with Haus's Oscillator

In 1978, Haus, et al. designed a reflection pulse oscillator, where a pulse travels back and forth on a transmission line[6,7] through reflection at both ends of the line [see Fig. 10.15(a)] [64]. The reflection soliton oscillator is an alteration of Haus's reflection pulse oscillator. This section compares the two oscillators in terms of circuit design style and pulse sharpening capability.

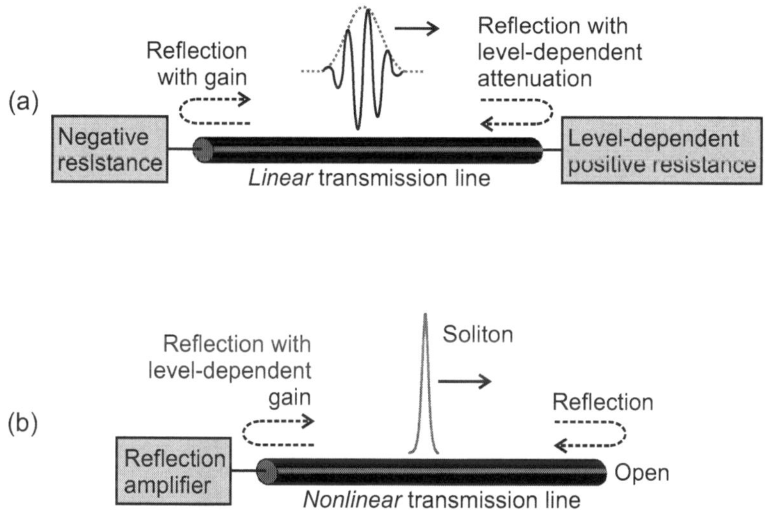

FIGURE 10.15
(a) Haus's reflection linear pulse oscillator [64]. (b) The reflection soliton pulse oscillator [redrawing of Fig. 10.1(b)].

Circuit design: Haus implemented the level-dependent gain using two separate active circuits [see Fig. 10.15(a)]: an IMPATT diode circuit terminating one end of the linear transmission line produces a nearly con-

[6] Although Haus, et al. used a metallic waveguide, it may be considered as a transmission line, as both have the same goal of introducing a delay.

[7] The actual pulse in Haus's oscillator is not a baseband pulse, but a pulse with a sinusoidal modulation as shown in Fig. 10.15(a). The essence of Haus's work, however, lies in the generation of the pulse envelope, as described here.

stant negative resistance (constant gain); a Schottky diode circuit terminating the other line end executes level-dependent attenuation. Their combined effect is the level-dependent gain. The adaptive bias scheme was incorporated in the Schottky diode circuit. In contrast, in the reflection soliton oscillator (see Fig. 10.9), the level-dependent gain and the adaptive bias scheme are all incorporated in a single one-port amplifier using bipolar transistors, which terminates one end of the NLTL (the other end of the NLTL is open) as shown in Fig. 10.15(b).

Pulse sharpening: In Haus's oscillator, pulse sharpening is provided only through the level-dependent gain, and thus, it is limited by the bandwidth of the active circuits providing the level-dependent gain. In the reflection soliton oscillator, the dominant pulse sharpening mechanism is provided by the NLTL. Since the NLTL's pulse sharpening (see Fig. 10.14) is not limited by the bandwidth of the reflection amplifier, this oscillator has an improved pulse sharpening ability. To experimentally show this improvement, a Haus oscillator was constructed by replacing varactors in the NLTL with linear capacitors. Figure 10.16(a) is the measured steady-state signal at the open end of the linear line in the Haus oscillator constructed for comparison, while Fig. 10.16(b) [redrawing of Fig. 10.12(a)] is the measured steady-state signal at the open end in the reflection soliton oscillator, which achieves a 5 times sharper pulsewidth.

10.5 Summary

This chapter presented the reflection soliton oscillator. Instead of the closed-loop nonlinear transmission line (NLTL) of the circular soliton oscillator presented in the previous chapter, this circuit consists of a straight NLTL, one end of which is connected to a one-port amplifier and the other end is open. The oscillation self-starts from ambient noise and in the steady state, the circuit self-generates a periodic train of short-duration pulses. While the one-port amplifier produces gain and stabilizes the oscillation, the NLTL introduces delay between two pulse events and substantially sharpens the pulse. The operation of the oscillator was confirmed by measurement.

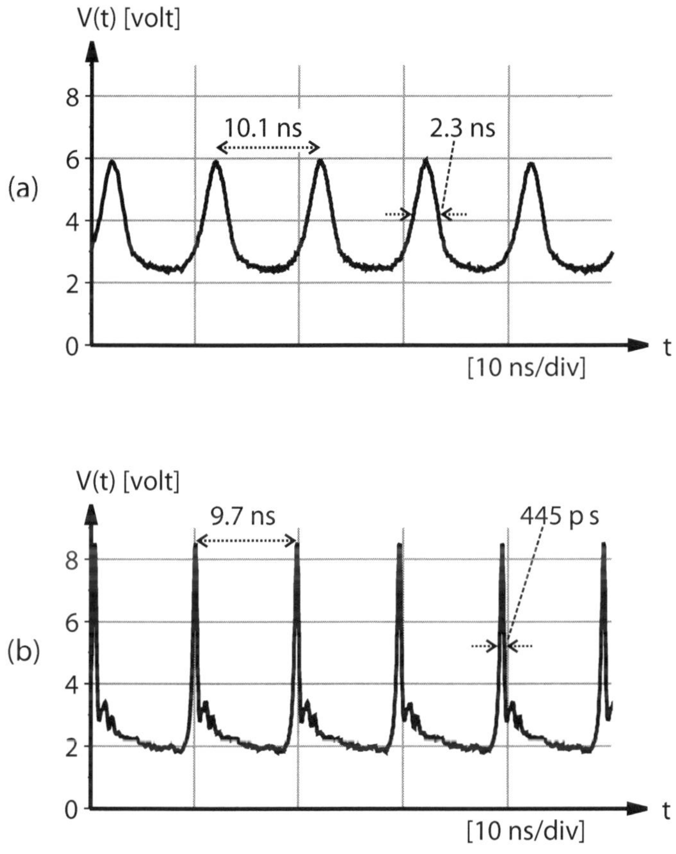

FIGURE 10.16
(a) Measured steady-state oscillation in a Haus oscillator built for comparison. (b) Measured steady-state oscillation in the reflection soliton oscillator [redrawing of Fig. 10.12(a)].

11

Chaotic Soliton Oscillator and Chaotic Communications

O. Ozgur Yildirim
Harvard University

Nan Sun
Harvard University

Xiaofeng Li
Harvard University

CONTENTS

Previous chapters of this book have focused on stable soliton oscillation by controlling the unruly soliton dynamics. In this chapter, however, we look at the opposite end of the spectrum, and try to attain an extreme oscillation instability, or chaos, by encouraging these unruly soliton dynamics. In both of the circular and reflection soliton oscillators [16, 68] described in the previous chapters, the stability of soliton oscillation is ensured by the level-dependent gain capability of the amplifiers used, that is, the amplifier provides gain for a soliton pulse higher than a particular threshold voltage, and attenuates small soliton pulses and perturbations. Hence only the highest soliton can survive in the system, and the circuit generates a stable train of solitons with constant amplitude and phase.

What happens if a full-gain amplifier is used instead of the amplifier with level-dependent gain in a soliton oscillator? In this way the amplifier would amplify any soliton pulse and perturbation generated in the NLTL. Shortly after, we would expect to see a number of solitons with different amplitudes, hence propagation speed (remember the amplitude-dependent speed property of solitons) on the NLTL, which would collide with each other, and further change their phases, or positions (remember the phase modulation after soliton collisions). Therefore, it might be possible to attain an unstable oscillation, which continually changes the phase and amplitude, using this topology. In this chapter, we will describe such a chaotic soliton oscillator. After describing the oscillator, we will show the chaotic behavior of it by simulation and Lyapunov analysis. Then, we will describe one possible application of the chaotic soliton oscillator, namely secure binary communication.

11.1 Chaos and Chaotic Communications

Chaos is mathematically defined as randomness generated by deterministic systems [69]. The sensitivity of chaotic systems to the initial conditions is the reason for this randomness, but because of the deterministic structure, this randomness follows a mathematical order. In other words, although chaotic systems behave in order, they are not predictable, because they are very sensitive to initial conditions and we mostly do not know all the information about the system.

This dual property of chaotic systems, i.e., randomness and deterministic order, can be explained by this example: Assume we have two identical chaotic systems with exactly the same initial conditions. In an ideal case, e.g., without noise, these systems will generate the same output no matter how long we wait because chaos is deterministic. However, the output of the systems is difficult to predict, because of the randomness. When the initial conditions are not exactly same for these identical chaotic systems, the two chaotic systems start to generate different outputs and the outputs' trajectories in phase space separate from each other.

The Lyapunov exponent is an indication of the rate of this separation and shows the sensitivity of the system to initial conditions. Quantitatively, two trajectories in phase space with initial separation $\triangle(0)$ diverge from each other

$$|\triangle(t)| \approx e^{\lambda t}|\triangle(0)| \tag{11.1}$$

where λ is the Lyapunov exponent. A system can have different Lyapunov exponents for different orientations. In general, if a system has at least one positive Lyapunov exponent in phase space, this will be a sufficient indication that it is a chaotic system [69].

Chaotic oscillators self-generate a chaotic signal. These circuits are

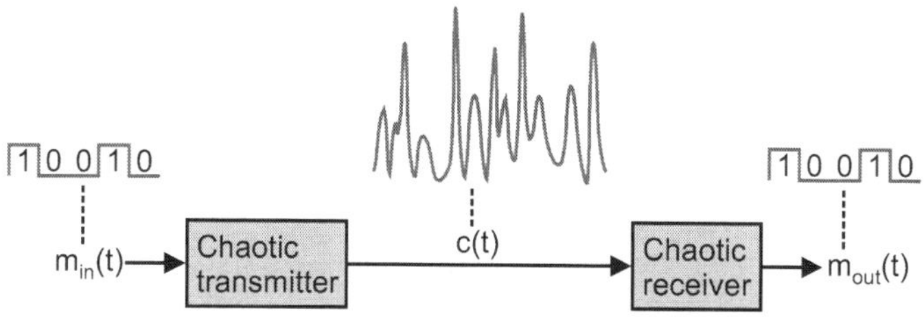

FIGURE 11.1
Basic concept of chaotic communication. Input message $m_{in}(t)$ is masked and carried by the chaotic carrier signal $c(t)$.

useful in physical random number generation [70], and in broadband and/or secure communication [71–73] as the carrier signal in broadband communication, and as both the carrier and masking signal in broadband secure communication. Broadband secure chaotic communication is an interesting application of chaotic soliton oscillator and we will detail its use in chaotic communications in Sec. 11.4.

Fig. 11.1 shows block diagram of a chaotic communication example. Input message $m_{in}(t)$ is modulated in the chaotic carrier signal $c(t)$ in the transmitter circuit, and the message output $m_{out}(t)$ is recovered in the receiver circuit from $c(t)$. Because of the randomness of the chaotic signal $c(t)$, it is very difficult for an outsider to get the message from $c(t)$. However, when the receiver is synchronized or matched to the transmitter, it is possible to get the data in the receiver because of the deterministic order of the chaotic signal $c(t)$.

There exist other secure communication systems, which are executed by algorithms, such as software-based classical [74] and quantum cryptographic systems [75]. On the contrary, communication with chaos is a hardware-based secure communication system, and it can be used alone or with the above software-based methods to enhance the security of the transmitted data.

As stated in previous chapters, the NLTL has a superb pulse sharpening ability. Therefore, the chaotic soliton oscillators potentially has the capability of generating an extremely broadband chaotic signal, which extends into the THz region [34, 76]. This directly corresponds to a high data rate in secure chaotic communication, making the chaotic soliton oscillator a candidate circuit for future secure communication applications.

11.2 Chaotic Soliton Oscillator

Let's recall a key property of solitons to understand the mechanism behind chaos. In Fig. 11.2, two different cases are compared. In the first case [see Fig. 11.2(a)], a single tall soliton propagates on an NLTL. In the second one [see Fig. 11.2(b)], there is an additional short soliton in front of the identical tall one. Because of the amplitude-dependent speed, the tall soliton will catch up to the short one and they will make a nonlinear collision. During the collision the amplitude modulates, and after the collision the phase modulates. That is, the tall soliton will end up at a different position than in the case without any collision. The short soliton also experiences a phase shift after the collision, which is not marked on the figure. The sign and amount of the phase shifts depend on the amplitude of each colliding soliton and their amplitude ratio [10]. If a soliton makes numerous collisions, then the pulse will end up in a totally different place than the initial expectation without any collisions. To attain a chaotic signal may be possible with numerous soliton collisions, where the phase and amplitude of the oscillation continually modulate, as in a chaotic signal.

Therefore, the chaotic soliton oscillator should encourage soliton collisions. One way to construct such a circuit is to place an NLTL and an amplifier in a closed-loop circular configuration [see Fig. 11.3(a)]. If the amplifier provides gain to all signal levels, any perturbation would be amplified and many different-sized solitons will form on the NLTL. Then, they will be again amplified by the amplifier and all these different-sized solitons will survive. In Fig. 11.3(a), a scenario is depicted where the amplifier amplifies both incoming tall and short solitons. After the amplification, since they propagate at different speeds (because of amplitude-dependent speed of solitons), they eventually collide. Actually, many solitons would be created by a system like these two, and they all circulate in the loop at different speeds, continually colliding with each other. The result would be a chaotic signal as depicted in Fig. 11.3(b). Otherwise, if the amplifier in Fig. 11.3(a) provided a level-dependent gain, it would eliminate the short soliton pulse shown in the figure, preventing the collision and preserving the stability.

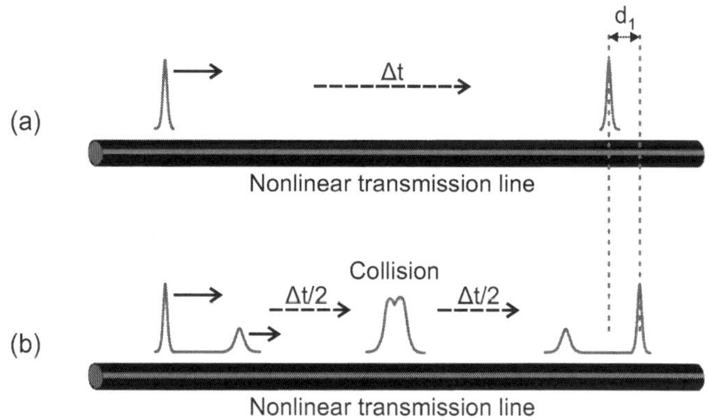

FIGURE 11.2
Depiction of a soliton collision and the effect of collision on amplitude and phase modulations.

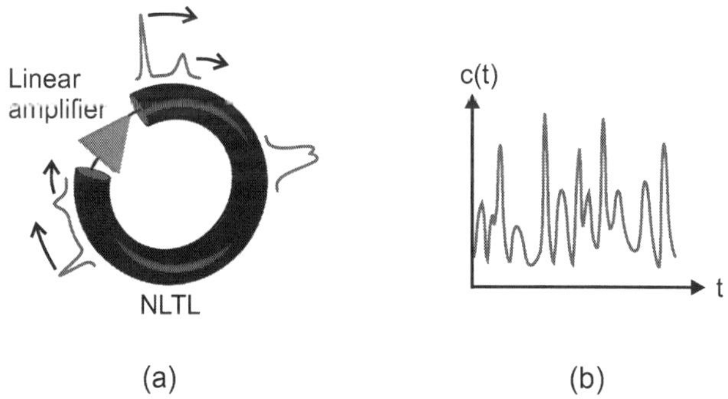

FIGURE 11.3
(a) Block diagram of the chaotic soliton oscillator. (b) Waveform depiction on one of the nodes of the NLTL.

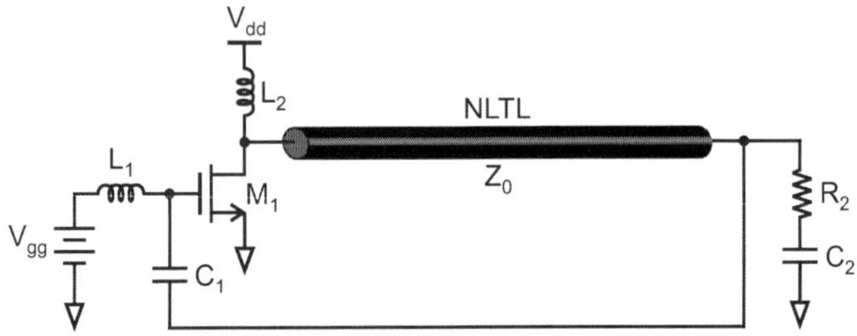

FIGURE 11.4
Schematic of the chaotic soliton oscillator used in the simulations.

11.3 Simulation of the Chaotic Soliton Oscillator

To demonstrate chaotic oscillation, we consider the circuit of Fig. 11.4. An amplifier and an NLTL are placed in a circular configuration to form a closed-loop. Here, we use an inverting amplifier for the sake of simplicity, although it does not matter whether the amplifier is inverting or noninverting, as long as it provides gain to any signal level. L_1 and C_1 have large values and work together as a bias tee. L_1 isolates the AC signal from the gate bias V_{gg}, and C_1 isolates the *dc* bias voltages of the gate and the NLTL. The NLTL is terminated with R_2-C_2 network, where R_2 is equal to the nominal characteristic impedance of the NLTL Z_0, and C_2 is the DC-blocking capacitor. Although in the chaotic soliton oscillator, a termination circuit is not a must, because we do not need to suppress reflections from the end of the NLTL to preserve the stability, we use the termination network to ensure a one-way signal propagation on the NLTL, which is to be used in the chaotic communication circuit of the next section.

We constructed this circuit in transistor-level in Cadence program. We used TSMC 0.18-μm mixed-signal CMOS technology. The NLTL consists of 20 LC sections. Each L has 50 nH inductance with 1 Ω series resistance. For varactors we use MOS transistors, where source and drain are tied to make a varactor diode. The nominal capacitance is chosen as 20 pF, resulting in a nominal 50 Ω characteristic impedance for the NLTL.

Fig. 11.5(a) shows simulated voltage waveform at the middle of the

NLTL, and Fig. 11.5(b) shows its Fourier transform. The unstable non-periodic waveform in the time domain, and the broad continuous spectral distribution both hint a chaotic oscillation.

Fig. 11.6 shows the oscillation trajectory in the (V(t), dV(t)/dt) phase space, where the trajectory fills what appears to be a bounded region. This phase space dynamics once again indicates that the oscillation can be chaotic, because an ordered oscillation follows a circular trajectory rather than filling the space.

As explained in the beginning of this chapter, a positive Lyapunov exponent is an indication of chaos. We simulated the Lyapunov exponent of our circuit as follows. We run two simulations with very close initial conditions and record all the twenty node voltages on the NLTL. We calculate the evolution of the absolute difference $[\log(\triangle(t))]$ between the NLTL node voltages of the two simulations as time progress. As shown in Fig. 11.7, $\log(\triangle(t))$ increases linearly with time, confirming that $\triangle(t)$ exponentially grows and the oscillation is indeed chaotic.

Fig. 11.8 shows the separation of two very close trajectories on the phase space plane. The node voltage in the middle of the NLTL is simulated twice with very close initial conditions. Because of the chaotic behavior of the system, these two initially very close points separate, and their final positions are distant from each other.

11.4 Simulation of Chaotic Binary Communication

One method to demonstrate transmission of data using a chaotic soliton oscillator is to use chaotic binary communication, which was previously implemented by using the Chua's chaotic oscillator [72] and the Lorenz-based chaotic oscillator [73]. In both of these implementations, the data transmission rate is very low, in the order of Kbps, because the speed of these systems is limited by the bandwidth of the operational amplifiers used in the Chua's resistor and the Lorenz-based chaos generator [72,73]. We applied the same communication principle to our system, which can have a much higher data transmission rate, owing to the superior broad bandwidth of the chaotic soliton oscillator.

Fig. 11.9 shows the detailed schematic of our chaotic binary communication system. While the transmitter is a closed-loop chaotic soliton oscillator, the receiver is an open loop circuit and driven by the transmitted signal. Binary data, $m_{in}(t)$, is modulated into the output of the transmitter, $c(t)$. The receiver is matched to the transmitter when $m_{in}(t)$ is logic 1, and then $c'(t) \approx c(t)$, and the error signal, $e(t) \approx 0$.

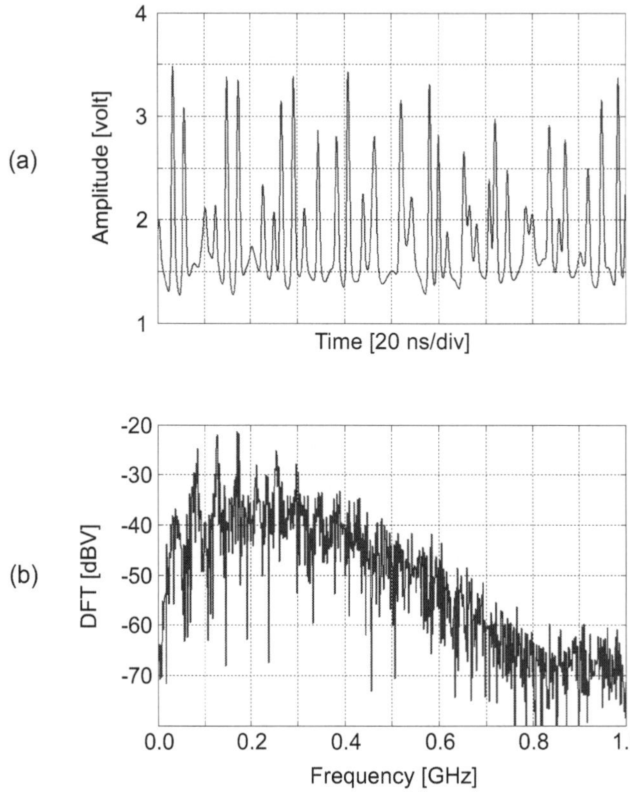

FIGURE 11.5
(a) Simulated output of the circuit. (b) Calculated spectrum of the above simulated waveform.

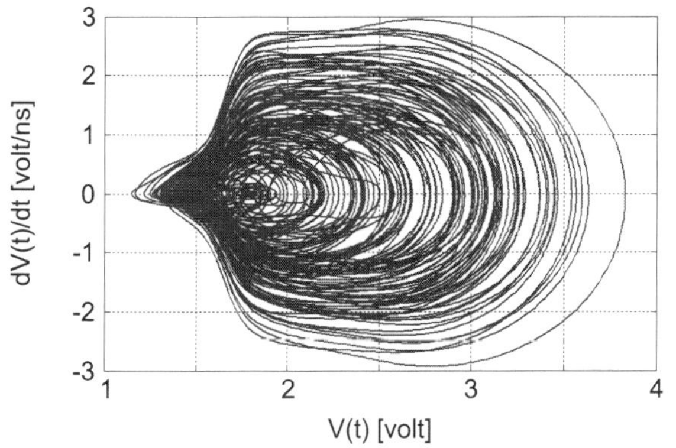

FIGURE 11.6
Phase space dynamics of the circuit. The trajectory fills what appears to be a bounded region.

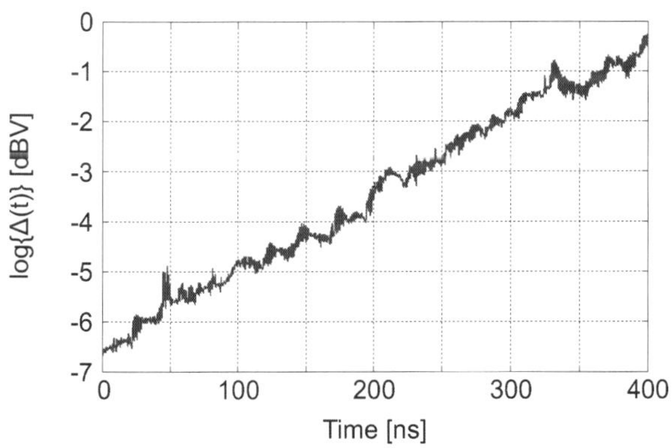

FIGURE 11.7
Simulated phase difference, indicating a positive Lyapunov exponent, and hence chaos.

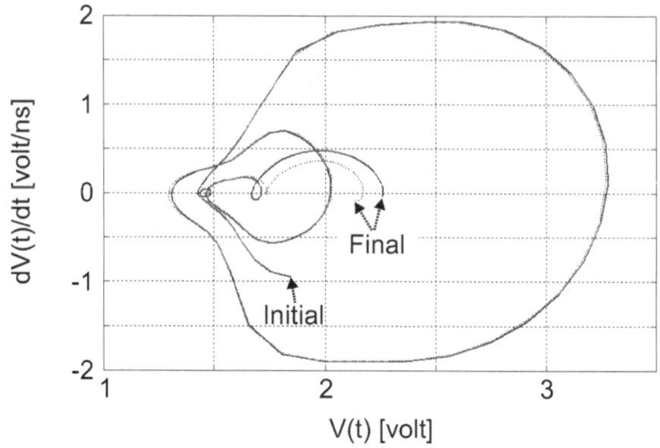

FIGURE 11.8
Separation of two very close trajectories in the phase space plane.

When $m_{in}(t)$ is logic 0; the receiver is mismatched to the transmitter, and $c'(t)$ is different from $c(t)$, and $e(t)$ has a significant signal power. Processing $e(t)$ by first an inverting amplifier, and then a rectifier, and finally another inverting amplifier, the output of the receiver, $m_{out}(t)$, can follow $m_{in}(t)$.

The transmitter is the same as the chaotic soliton oscillator of Section 11.3 except that the transistor is M_2 and the inductor L_3. When $m_{in}(t)$ is logic 1, M_2 is OFF, matching with the receiver circuit. When $m_{in}(t)$ is logic 0, M_2 is ON, changing the load inductance of M_1, mismatching to the receiver circuit. The voltage on the termination is the signal to be transmitted. In the simulation, we used an ideal wire to send the information. In the practical cases, the signal can be sent by a transmission line, or a wideband antenna, both of which have an input impedance. In these cases, the input impedance of these media would be the termination, and no separate termination would be necessary. The transmitted signal $c(t)$ derives the input of the open loop circuit in the receiver, and another chaotic signal $c'(t)$ is generated on the termination of the receiver. $c'(t)$ matches to $c(t)$, depending on the value of the input message $m_{in}(t)$. A balun works as a subtracter, such that $e(t) = c'(t) - c(t)$. The balun generates a nonzero error voltage $e(t)$ when there is mismatch between the transmitter and receiver, but an almost zero error voltage $e(t)$ when they match. The reasons for not being exactly zero are noise, not perfectly matched components, and loading

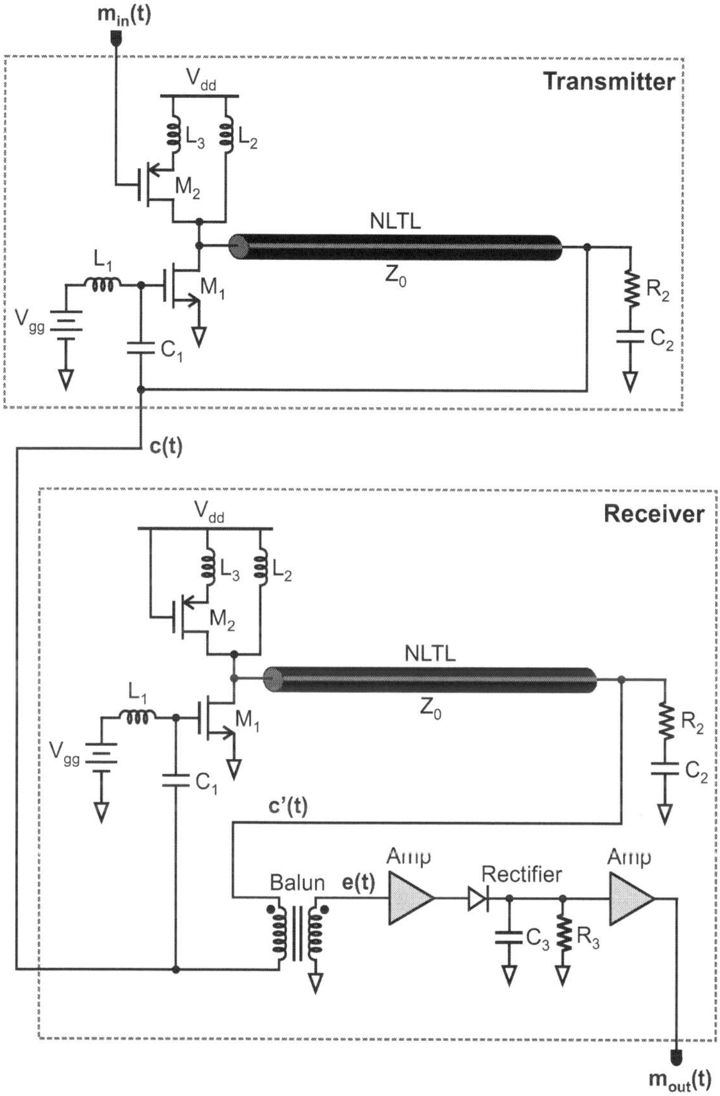

FIGURE 11.9
The binary chaotic soliton communication system. While the transmitter is a closed-loop chaotic soliton oscillator, the receiver is an open loop circuit and is driven by the transmitted signal.

of components. An inverting amplifier is used to increase voltage levels before the rectification, because the diode in the rectifier has around 0.7-V threshold voltage, which is higher than the balun output. The output of this amplifier is rectified, and low pass filtered by the R_3-C_3 network. This voltage is again amplified by an inverting amplifier to alter the polarity of the signal, so that $m_{out}(t)$ follows the same polarity as $m_{in}(t)$.

Fig. 11.10 shows the Cadence simulation results of this chaotic communication system. $m_{in}(t)$ is a 20-Mbps signal (50-ns bit duration). In the rectifier the R_3C_3 time constant is selected 25 ns, a comparable time with the data duration, 50 ns. A too short time constant would not eliminate the ripples of the balun's output, and a too large time constant would fail following the speed of the input signal.

For the reliability of the above secure communication mechanism, it is required that to get the message data without a matched receiver circuit is difficult. When we look at the time-domain transmitted signal [see $c(t)$ in Fig. 11.10], the data modulation is undistinguishable. A more vigorous method to understand the effect of the message modulation on the carrier signal is to look at the spectrum. Fig. 11.11(a) shows the spectrum of the time-domain signal $c(t)$ of Fig. 11.10. The spectrum component at 10 MHz is undistinguishable (10 MHz corresponds to the 20-Mbps data transmission rate). To investigate the worst case, we apply a 10-MHz square wave as the input message (corresponding to 20-Mbps data of alternating logic 0's and logic 1's), and spectrum of $c(t)$ in such a case is shown in Fig. 11.11(b). Now, the message signal at 10 MHz is distinguishable, but it is 22 dB weaker than the peak spectrum component, making it difficult to be captured by an outsider. Moreover, a higher data rate may be completely embedded in the chaotic spectrum, improving the security of data transmission.

11.5 Summary

The chaotic soliton oscillator generates a broadband chaotic output by promoting nonlinear collisions of solitons. The oscillator, confirmed by Cadence simulation, has a positive Lyapunov exponent, indicating that the circuit is chaotic. The chaotic soliton oscillator can be useful as a physical random number generator and as building blocks of physical secure communications. Indeed, an example circuit, a binary chaotic communication system, was described by using the chaotic soliton oscillator, and its operation was simulated by Cadence program.

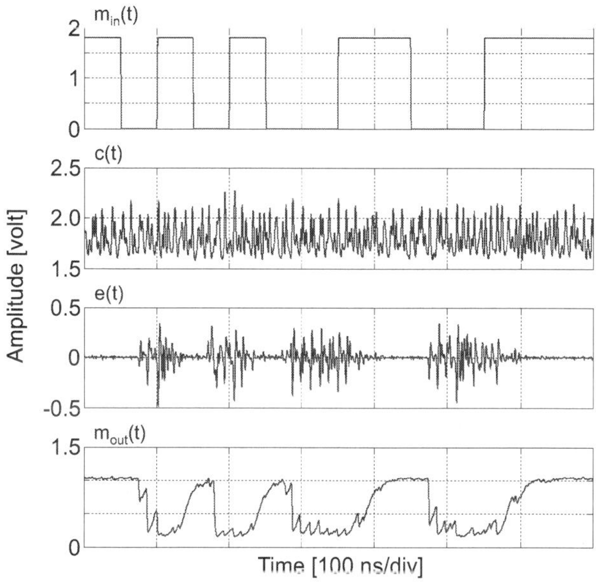

FIGURE 11.10
Simulated waveforms of the binary chaotic soliton communication.

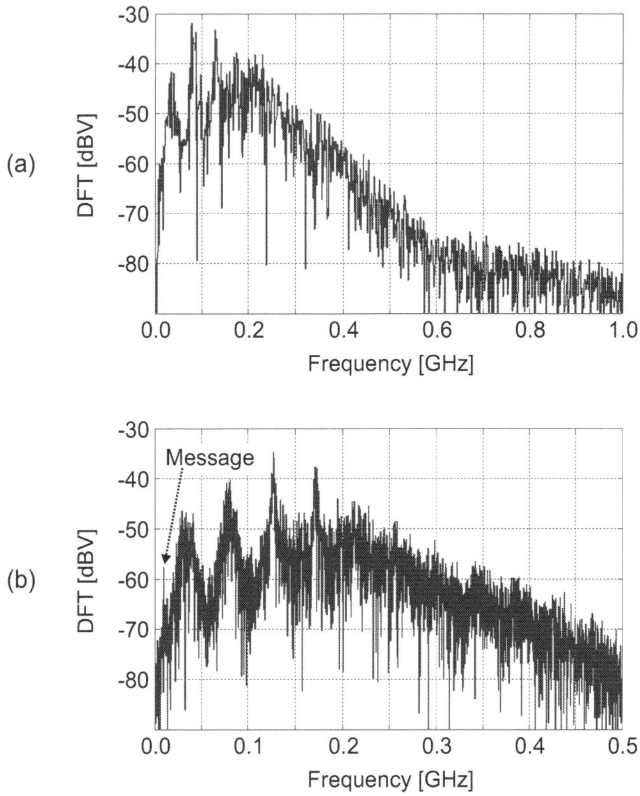

FIGURE 11.11
(a) Spectrum of $c(t)$ of Fig. 11.10. The message input is completely masked in the spectrum. (b) Spectrum of $c(t)$ when $m_{in}(t)$ is a 10-MHz square wave, examining the worst case. The message input is now visible in the spectrum, but it is still very small.

12

Phase Noise of Soliton Oscillators

Xiaofeng Li
Harvard University

CONTENTS

In this chapter we present phase noise analyses of the electrical soliton oscillators introduced in Chapter 9 and 10. We will start with a brief review of the fundamental phase noise theory, and then examine the particular phase noise process in soliton oscillators, including contributions from direct phase perturbations and indirect phase errors caused by amplitude-to-phase conversion. In addition, we will compare, by both analytical calculations and experiments, the phase noise of soliton oscillator with other oscillators based on linear transmission lines, and discuss key design strategies for minimizing the phase noise of soliton oscillators.

12.1 Phase Noise Fundamentals

Phase noise in self-sustained oscillators causes timing inaccuracy or frequency instability, and gives rise to spectral linewidth broadening of oscillator's output. It is among the most essential and interesting aspects of oscillator's dynamics and quality, and has been extensively studied since 1960s in the context of lasers (optical oscillator) and electrical oscillators [77–79].

The essence of phase noise may be mostly easily understood using an LC oscillator [Fig. 12.1(a)] as example [77]. The voltage V across, and the current I in, the LC tank, represent the oscillator's state. The steady-state oscillation follows a closed-loop trajectory, or limit cycle, in the 2-dimensional (2-D) V-I state space [Fig. 12.1(b)]. Noise perturbs the oscillation, causing amplitude and phase errors. The amplitude error that puts oscillation *off* the limit cycle is constantly corrected by the oscillator's tendency to return to its limit cycle. In contrast, the phase error *on* the limit cycle along its tangential direction accumulates without bound, for no mechanism to reset the phase exists. In other words, the phase undergoes a diffusion process along the limit cycle. In time domain, this *phase diffusion* causes timing variations, while in frequency domain it gives rise to frequency instability. The oscillator's output spectrum is thus broadened around the oscillation frequency [Fig. 12.1(c)], which is known as phase noise.

Mathematically, in the presence of only white noise, the phase diffusion so occurs that the variance of the phase ϕ of the oscillation, $\cos(\omega_0 t + \phi)$, grows linearly with time

$$\langle \phi^2 \rangle = 2Dt, \tag{12.1}$$

where D is the phase diffusion rate. This leads to the well-known Lorentzian phase noise

$$\mathcal{L}\{\Delta\omega\} = \frac{2D}{(\Delta\omega)^2 + D^2} \tag{12.2}$$

and the f^{-2} behavior for $\Delta\omega \gg D$:

$$\mathcal{L}\{\Delta\omega\} \approx \frac{2D}{(\Delta\omega)^2}, \tag{12.3}$$

which is familiar from D. Leeson's paper [80].

As seen, once the phase diffusion rate D is determined, phase noise is known. D can be determined by evaluating the phase error along

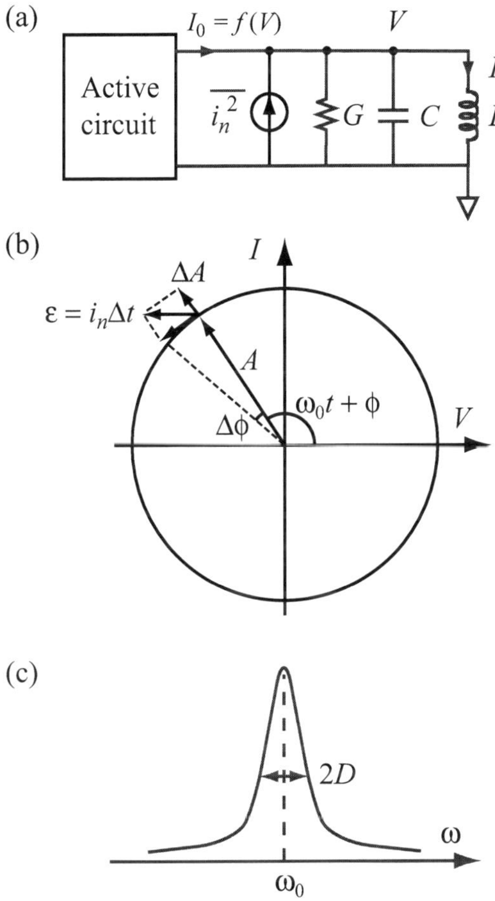

FIGURE 12.1
(a) Circuit diagram of a self-sustained LC oscillator where the voltage V on capacitor C is perturbed by the thermal noise i_n generated from parallel loss G. (b) Limit cycle in phase space. The phase error $\Delta\phi$ is obtained by projecting the perturbation ε along the tangential direction of the limit cycle, which depends on where ε occurs on the limit cycle. (c) The oscillator's output spectrum is broadened around its oscillation frequency due to phase noise.

the tangential direction of the limit cycle for a given noise perturbation. As illustrated in Fig. 12.1(b), the phase error depends not only on the noise level, but also on the oscillator state, or the limit cycle position where the perturbation occurs [77]. Therefore, this state-dependent, or, time-varying phase error is averaged along the limit cycle, thus, D will be a function of both noise level and oscillator's waveform (limit cycle shape) [77]. This time-variant property has been greatly exploited by the circuit community for low noise oscillator design [81].

While it has been well understood for LC oscillators, phase noise processes are more involved in soliton oscillators. The difficulty arises from two aspects. First, soliton oscillators are distributed oscillators, or wave-based oscillators, where energy storage components are distributed along waveguides or transmission lines to propagate electromagnetic waves. The oscillating signal is an electromagnetic wave (pulse-shaped in the case of soliton oscillators) represented by a large number of voltage and current variables distributed along the transmission lines. It is difficult to visualize how these distributed voltage/current variables are *collectively* disturbed by noise in space and time. Second, the special nonlinear dynamics of soliton pulse makes the problem even more complicated. These two aspects are treated separately in the next two sections.

Our starting point is the mathematical framework established in 1989 by F. Kaertner [78], where he extended the phase noise study to deal with general oscillator systems whose states are represented by N voltage and current variables in N-dimensional (N-D) state space. Under Kaertner's framework, noise perturbations are decomposed into a component along the tangential direction of the oscillator's limit cycle in the N-D state space, and components orthogonal to the tangential direction. The former component corresponds to phase perturbation that directly drives the phase diffusion process. Besides, Kaertner considered how the latter components, corresponding to amplitude perturbations, indirectly contribute to phase diffusion through amplitude-to-phase noise conversion determined by the oscillator's dynamics. This calculation would again follow the limit cycle and average the state-dependent phase errors in determining D.

This general framework by Kaertner, including the orthogonal projection, is the basis of our phase noise analysis of soliton oscillators. To carry it out in full, however, requires diagonalizing large matrices corresponding to oscillator's dynamical equations linearized in proximity of the limit cycle. Instead, we translate Kaertner's mathematical language to a physically intuitive analysis that captures the essence of phase noise process in soliton oscillators. In addition, for mathematical simplicity, we will focus on phase noise incurred by white noise only, although the analysis can be extended to deal with $1/f$ noise sources as in [79].

12.2 Phase Noise Due to Direct Phase Perturbation

In this section, we calculate the phase noise driven directly by the tangential projections of noise perturbations. The analysis is not limited to soliton oscillators, but applicable to general distributed oscillators with arbitrary shape of waveform (e.g., sinusoidal standing-wave oscillators and linear pulse oscillators). Suppose that the transmission line is an artificial one and consists of N LC sections (Fig. 12.2). The corresponding oscillator has $2N$ state variables, each corresponding to a voltage or current variable of a capacitor or inductor, and the oscillator state is represented by a vector $\vec{U} = (\{V_n\}, \{I_n\})$ $(n = 1, 2, ..., N)$ in the $2N$-D state space[1]. In steady state, the state vector traces a closed-loop trajectory, or limit cycle, $\vec{U}_s(t)$, in the $2N$-D space (Fig. 12.3).

Noise that perturbs the oscillator's state, $\vec{U}$, or in other words, the oscillator's limit cycle position, can be distributed along the line (e.g., thermal noise associated with the distributed loss in the line) or lumped (e.g., noise from the lumped active circuit and the termination of transmission line). We will treat these two noise sources separately in the following two subsections.

12.2.1 Distributed Noise Sources

We first consider the distributed noise from an artificial transmission line with $2N$ state variables (Fig. 12.2). The calculation of the direct phase perturbation by the distributed noise runs in two steps. First, by analyzing the oscillator dynamics in the presence of the noise, we identify the vector representing the noise perturbation. Second, we project this noise perturbation vector onto the tangential direction of the limit cycle to calculate the phase error.

To identify the vector of noise perturbation in the $2N$-D state space,

[1] It is assumed that the oscillator's active circuit is memoryless compared to the transmission line, not generating extra state variables.

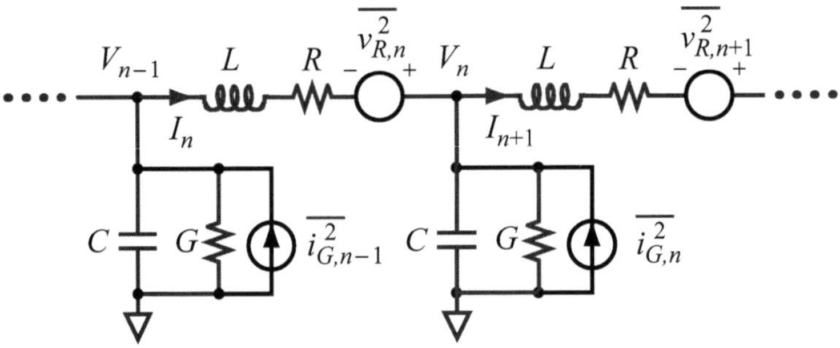

FIGURE 12.2
Artificial transmission line consisting of N LC-sections. $n = 1, 2, ..., N$. Loss components R and G are included with associate noise.

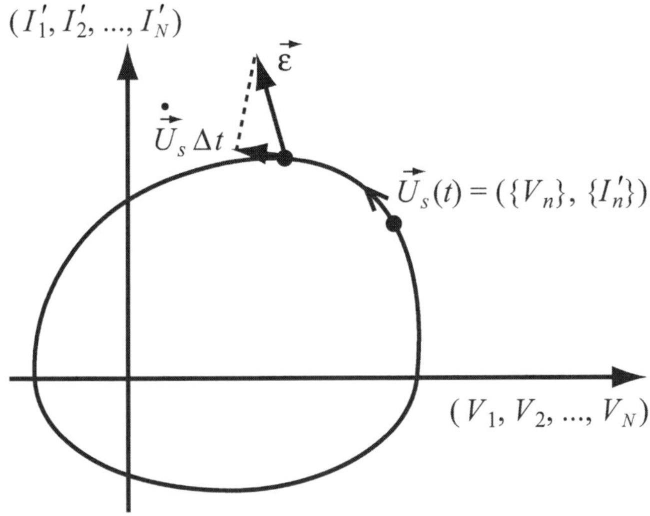

FIGURE 12.3
Oscillator's limit cycle in $2N$-D state space. For illustration purpose, we only show two axes.

we apply Kirchhoff's law in Fig. 12.2:[2]

$$C\dot{V}_n = I_n - I_{n+1} - V_n G + i_{G,n}(t), \tag{12.4a}$$
$$L\dot{I}_n = V_{n-1} - V_n - I_n R + v_{R,n}(t), \tag{12.4b}$$

where $n = 1, 2, ..., N$ and the dot above a variable represents a time derivative. Resistance R and conductance G represent loss in the line. Their associated Nyquist voltage and current thermal noise, $v_{R,n}$ and $i_{G,n}$, satisfy $\overline{v_{R,n}^2}/\Delta f = 4kTR$ and $\overline{i_{G,n}^2}/\Delta f = 4kTG$.

Change of variables, $I'_n \equiv I_n Z_0$, $R' \equiv R/Z_0$, and $G' \equiv GZ_0$, with an impedance-unit parameter $Z_0 \equiv \sqrt{L/C}$, renders R' and G' dimensionless, and I'_n with the dimension of voltage[3]. With the new variables, we rewrite (12.4) into:

$$dV_n = [(I'_n - I'_{n+1} - V_n G')dt + \varepsilon_{G,n}]/\sqrt{LC}, \tag{12.5a}$$
$$dI'_n = [(V_{n-1} - V_n - I'_n R')dt + \varepsilon_{R,n}]/\sqrt{LC}, \tag{12.5b}$$

where $\varepsilon_{R,n} \equiv v_{R,n}(t)dt$ and $\varepsilon_{G,n} \equiv Z_0 i_{G,n}(t)dt$ are rescaled noise terms that are independent Gaussian random variables with zero mean and variance $\langle \varepsilon_{R,n}^2 \rangle = 2kTR'Z_0 dt$, $\langle \varepsilon_{G,n}^2 \rangle = 2kTG'Z_0 dt$ [82]. Eq. (12.5) identifies a random vector $\vec{\varepsilon} = (\{\varepsilon_{G,n}\}, \{\varepsilon_{R,n}\})/\sqrt{LC}$ as the noise perturbation vector during the time interval $[t, t+dt]$. Also note that Eq. (12.5) treats voltage and current variables symmetrically, or, in the equal unit, thanks to the change of variable, $I'_n \equiv I_n Z_0$.

We now project $\vec{\varepsilon}$ onto the tangential direction of the limit cycle, or the direction of motion in the state space, which is defined by the state-space velocity $\dot{\vec{U}}_s$ (Fig. 12.3). The inner product between $\vec{\varepsilon}$ and the unit vector $\dot{\vec{U}}_s/|\dot{\vec{U}}_s|$ is the perturbation along the direction of motion. Division of this tangential perturbation by the magnitude of the velocity $|\dot{\vec{U}}_s|$ yields the timing error, Δt, which is converted to the phase error ϕ using $\phi = \omega_0 \Delta t$, where ω_0 is the oscillation frequency. In sum,

$$\phi = \omega_0 \frac{\vec{\varepsilon} \cdot \dot{\vec{U}}_s}{|\dot{\vec{U}}_s|^2} = \frac{\omega_0}{\sqrt{LC}} \frac{\sum_n \varepsilon_{G,n} \dot{V}_n + \sum_n \varepsilon_{R,n} \dot{I}'_n}{\sum_n \dot{V}_n^{\,2} + \sum_n \dot{I}'^{\,2}_n}. \tag{12.6}$$

Since $\varepsilon_{R,n}$ and $\varepsilon_{G,n}$ are independent Gaussian random variables with

[2]In case of the nonlinear transmission line where C is voltage dependent, the nominal C-value is used in the following calculations for approximation.

[3]This transformation is exact for linear transmission lines, but an approximation in case of the nonlinear transmission line where C is voltage dependent. The nominal C-value and line impedance Z_0 are used in the following calculations for simplicity.

zero mean, ϕ is also Gaussian with zero mean, and its variance is given by:

$$\langle \phi^2 \rangle = 2 \times \frac{kTZ_0\omega_0^2}{LC} \frac{G' \sum_n \dot{V}_n^{\ 2} + R' \sum_n \dot{I}_n'^{\ 2}}{\left[\sum_n \dot{V}_n^{\ 2} + \sum_n \dot{I}_n'^{\ 2}\right]^2} \times dt. \tag{12.7}$$

Since the phase error must accumulate, growing its variance linearly with time, (12.7) can be directly compared with the phase diffusion model (12.1) to find the phase diffusion rate D:

$$D(t) = \frac{kTZ_0\omega_0^2}{LC} \frac{G' \sum_n \dot{V}_n^{\ 2} + R' \sum_n \dot{I}_n'^{\ 2}}{\left[\sum_n \dot{V}_n^{\ 2} + \sum_n \dot{I}_n'^{\ 2}\right]^2}. \tag{12.8}$$

Since V_n and I_n are periodic functions of time, $D(t)$ above is an instantaneous rate at a given time, and it varies periodically with time. This is because the size of the phase error even for a fixed noise vector $\vec{\varepsilon}$ depends on the state of oscillation (where the state lies on the limit cycle), as indicated by (12.6) [77, 78]. The periodically-varying diffusion rate is usually time-averaged over one oscillation period to yield a constant diffusion rate, which matters over a long run:

$$D = \frac{kTZ_0\omega_0^2}{LC} \left\langle \frac{G' \sum_n \dot{V}_n^{\ 2} + R' \sum_n \dot{I}_n'^{\ 2}}{\left[\sum_n \dot{V}_n^{\ 2} + \sum_n \dot{I}_n'^{\ 2}\right]^2} \right\rangle, \tag{12.9}$$

where $\langle \cdot \rangle$ represents the time average. This concludes our calculation of the direct phase perturbation and the resulting phase diffusion rate, arising from the distributed noise in the artificial transmission line.

In the continuous limit where $2N \to \infty$ with infinitesimally small LC sections, the phase diffusion rate can be directly obtained from (12.9) by replacing the summations with integrals ($\sum_n \to (\Delta x)^{-1} \int dx$) and replacing L, C, R' and G' with their values per unit length, l, c, r' and g' (L, C, R', $G' \to l\Delta x$, $c\Delta x$, $r'\Delta x$, $g'\Delta x$):

$$D = kTZ_0 v^2 \omega_0^2 \left\langle \frac{g' \int (\partial_t V)^2 dx + r' \int (\partial_t I')^2 dx}{\left[\int (\partial_t V)^2 dx + \int (\partial_t I')^2 dx\right]^2} \right\rangle, \tag{12.10}$$

where x is the spatial coordinate along the line, ∂_t is partial time-derivative, and $v = 1/\sqrt{lc}$ is the wave velocity on the line (not to be confused with the state-space velocity $\dot{\vec{U}}_s$).

We now apply these results to the soliton oscillators and compare them to other types of oscillators (Fig. 12.4), including the LC oscillator, sinusoidal standing-wave oscillator, and reflection linear pulse oscillator. The reflection linear pulse oscillator [Fig. 12.4(b)] has essentially the same construction as the reflection soliton oscillator [Fig. 12.4(c)], but with the nonlinear transmission line replaced by a linear one. It also produces a periodic train of pulses, but with much wider pulse width. The standing-wave oscillator is built with a similar structure. Its active circuit, however, contains no saturable absorber, but is designed to amplify small voltages and attenuates large voltages as in standard LC oscillators. As a result, a sinusoidal $\lambda/2$ standing wave is formed on the linear transmission line [Fig. 12.4(a)]. We start with the simplest case of a lumped LC oscillator:

1. *Lumped LC oscillator*, Fig. 12.1(a):
   The voltage and current in the LC tank of an almost sinusoidal oscillator are given by $V(t) = A\cos\omega_0 t$ and $I'(t) = A\sin\omega_0 t$. Plugging these into (12.9) with $N = 1$ while only taking into account G with $R = 0$ to model a parallel loss in the tank, we obtain

   $$D_{LC} = \frac{1}{4}\frac{kT}{E_s}\frac{\omega_0}{Q}, \qquad (12.11)$$

   where $E_s = CA^2/2$ is the total oscillation energy stored in the tank and $Q = \omega_0 C/G$ is the tank's quality factor. This can be directly converted to the well-known Leeson's formula [80]. Note that kT is the total thermal energy in the tank (corresponding to noise), because L and C each stores a mean thermal energy of $kT/2$. The ratio of the thermal energy to the oscillation energy E_s is the noise-to-signal ratio.

2. *Sinusoidal standing wave oscillator*, Fig. 12.4(a):
   Voltage and current standing waves in a continuous transmission line are $V(x,t) = A\cos\beta x\cos\omega_0 t$ and $I'(x,t) = A\sin\beta x\sin\omega_0 t$ where $\beta = \omega_0/v$ is the wave number. Plugging these into (12.10) and performing the integrals over the line length d, we find

   $$D_{\mathrm{sin}} = \frac{kTZ_0v^2}{A^2d}(r' + g') = \frac{1}{4}\frac{kT}{E_s}\frac{\omega_0}{Q}, \qquad (12.12)$$

   where $E_s = cdA^2/4$ is the total energy stored in the transmission line, and $Q = \omega_0/(r/l + g/c)$ is the quality factor of the line. Note that the phase diffusion rate of the sinusoidal standing wave oscillator above conforms to that of the lumped LC oscillator in (12.11). This is expected and can be understood in the Fourier domain: although the distributed voltage and current variables in the transmission line correspond to a large (or infinite) number of degrees of freedom in general, in the sinusoidal standing wave

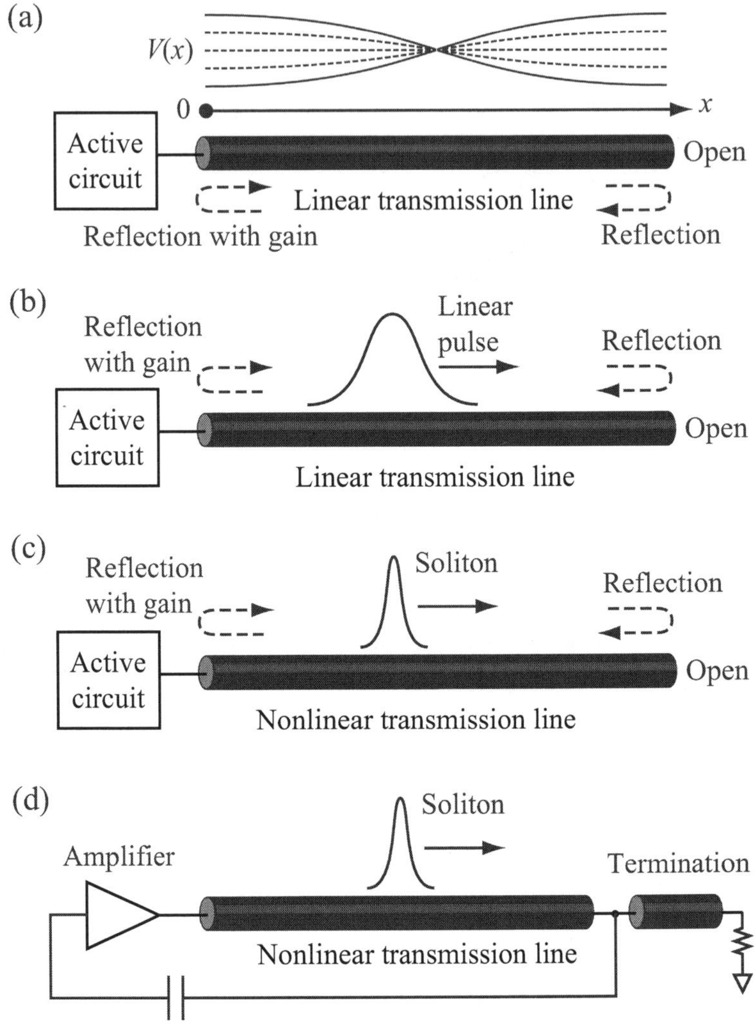

FIGURE 12.4
Distributed oscillator examples: (a) $\lambda/2$ sinusoidal standing-wave oscillator; (b) linear pulse oscillator; (c) reflection soliton oscillator; (d) ring soliton oscillator.

oscillator, only one resonance mode of the transmission line is effectively excited. Like an LC tank, this resonance mode possesses only two degrees of freedom, namely the electric and magnetic fields, each storing a mean thermal energy of $kT/2$. It is then possible to treat sinusoidal standing wave oscillators effectively as lumped LC oscillators. In fact, using the Fourier amplitudes as state variables, the dynamical equations of standing wave oscillators can be reduced to those of lumped LC oscillators. Such reduction was already carried out in the early stage of phase noise study for masers and lasers, whose microwave/optical cavities are indeed distributed waveguides that support microwave/optical standing waves [83]. However, this convenient thermodynamic argument in Fourier domain fails in the case of pulse oscillators, as we will now show.

3. *Reflection pulse (linear and soliton) oscillators*, Figs. 12.4(b)(c): The bell-shaped pulse in both linear and soliton reflection pulse oscillators can be described by the same functional form $V = A\text{sech}^2[B(x - vt)] \approx I'$ with a full spatial width at half maximum, $W \approx 1.76/B$, although the soliton pulse tends to be sharper (larger B). Therefore, their phase diffusion rates can be calculated together using (12.10). Since the pulse typically spans a length much shorter than the total length of the line, we can replace the integrals over the line length d in (12.10) with integrals over the entire x-axis[4], i.e.,

$$\int_0^d (\partial_t V)^2 dx \approx \int_{-\infty}^{\infty} (\partial_t V)^2 dx \approx 1.88 A^2 v^2 / W, \tag{12.13}$$

to obtain

$$D_{\text{pulse}} \approx \frac{kTZ_0\omega_0^2 W}{7.5A^2}(r' + g') \approx \frac{W^2}{d^2}\frac{kT}{E_s}\frac{\omega_0}{Q}, \tag{12.14}$$

where $E_s \approx 0.76cWA^2$ is the total energy carried by the pulse. Eq. (12.14) shows that phase noise of the pulse oscillator improves as the pulse gets sharper. This can be physically understood in the time domain: noise perturbation generated at position x affects the oscillator's phase or the timing of the pulse only when the pulse passes through the point. Its chance to

[4]When the pulse reflects at the line ends, the instantaneous phase diffusion rate $D(t)$ can be slightly modified due to superposition of the incident and reflected pulses. The difference, however, is minimal after time averaging.

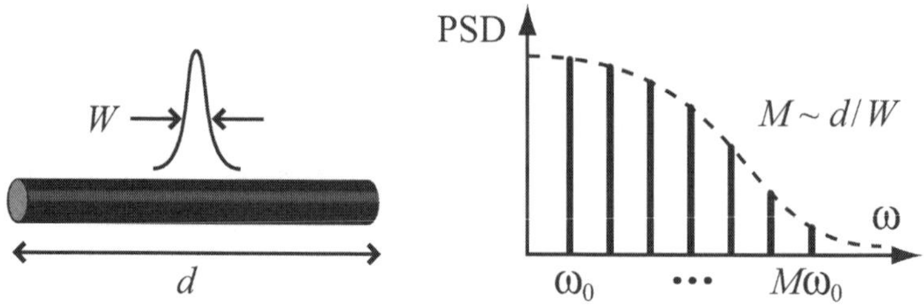

FIGURE 12.5
Power spectrum density (PSD) of periodic pulses

enter the phase noise process thus becomes smaller as the pulse width decreases, yielding lower D and better phase noise. Since $M = d/W$ is approximately the number of excited harmonic modes that constitute the pulse (Fig. 12.5), (12.14) may be expressed as $D \sim (1/M^2) \times (kT/E_s) \times (\omega_0/Q)$. Had we treated the problem in the Fourier domain assuming that the modes were independent of one another, each mode would store a thermal energy of kT, and $D \sim M \times (kT/E_s) \times (\omega_0/Q)$ would result. This, however, is an incorrect result, larger than the true D value above by a factor of M^3. The Fourier domain analysis fails because the harmonic modes are not independent, as their relative phases are coupled (or mode-locked) together. From (12.12) and (12.14),

$$\frac{D_{\text{pulse}}}{D_{\sin}} \approx 1.3\frac{W}{d}, \tag{12.15}$$

if the standing wave oscillator and the pulsed oscillator have the same amplitude for the same noise level, and

$$\frac{D_{\text{pulse}}}{D_{\sin}} \approx \frac{4W^2}{d^2}, \tag{12.16}$$

if the two oscillators have the same power dissipation for the same noise level. Therefore, for a very sharp pulse ($W \ll d$), $D_{\text{pulse}} \ll D_{\sin}$.

4. *Ring soliton oscillator*, Fig. 12.4(d):
As in the reflection soliton oscillator, the pulse in a ring soliton oscillator can be expressed by the same functional form $V = A\text{sech}^2[B(x - vt)] \approx I'$, which yields the same phase diffusion rate as in 12.14:

$$D_{\text{pulse}} \approx \frac{kTZ_0\omega_0^2 W}{7.5A^2}(r' + g'). \tag{12.17}$$

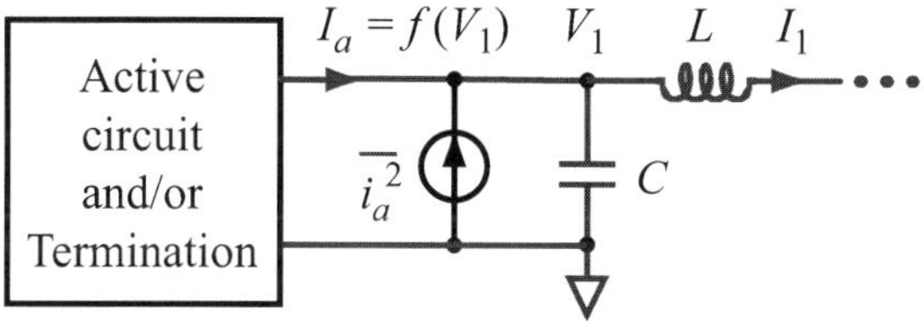

FIGURE 12.6
Circuit model for the active circuit with lumped noise.

The only difference is that the oscillation frequency ω_0 is doubled for the ring soliton oscillator as the round-trip distance is halved.

12.2.2 Lumped Noise Sources

The direct phase perturbation by the lumped noise from the active circuit and termination in Fig. 12.4 can be calculated in a similar fashion. At the end point of transmission line (Fig. 12.6), the Kirchhoff's law gives

$$C\dot{V}_1 = I_a - I_1 + i_a(t), \tag{12.18}$$

where $I_a = f(V_1)$ describes the characteristics of the active circuit/termination. Its associated noise source is lumped into $i_a(t)$ and assumed to be white Gaussian with autocorrelation $\langle i_a(t_1)i_a(t_2)\rangle = 2kTG_a(t_1)\delta(t_1 - t_2)$. The noise level, represented by the equivalent output noise conductance $G_a(t)$, is generally a function of time, which varies periodically with the oscillation [81].

As in Sec. 12.2.1, using change of variables $I'_n = Z_0 I_n$ and $G'_a = Z_0 G_a$, we rewrite (12.18) into:

$$dV_1 = [(I'_a - I'_1)dt + \varepsilon_a]/\sqrt{LC}, \tag{12.19}$$

where $\varepsilon_a \equiv Z_0 i_a(t)dt$ is a Gaussian random variable with zero mean and variance $\langle \varepsilon_a^2 \rangle = 2kTZ_0G'_a dt$. This equation identifies $\vec{\varepsilon} = (\varepsilon_a/\sqrt{LC}, 0, ..., 0)$ as a noise perturbation vector in the $2N$-D state space. By projecting $\vec{\varepsilon}$ along the tangential direction of the limit cycle and following the same procedure in Sec. 12.2.1, we obtain the following phase diffusion rate due to the direct phase perturbation by the lumped noise:

$$D = \frac{kTZ_0\omega_0^2}{LC}\left\langle \frac{G'_a(t)\dot{V}_1^2(t)}{\left[\sum_n \dot{V}_n{}^2 + \sum_n \dot{I}'_n{}^2\right]^2} \right\rangle, \tag{12.20}$$

In the continuous limit, it becomes

$$D = kTZ_0 v^2 \omega_0^2 \left\langle \frac{G_a'(t)\dot{V}^2(0,t)}{\left[\int (\partial_t V)^2 dx + \int (\partial_t I')^2 dx\right]^2} \right\rangle. \tag{12.21}$$

We again apply these results to the oscillators in Fig. 12.4. Evaluation of (12.20) and (12.21) requires knowledge of the specific form of the time-varying noise level $G_a'(t)$. In each of our experimental circuits, $G_a'(t)$ is almost a constant (Sec. 12.4.3). Therefore, a constant G_a' is used in the following calculations.

1. *Sinusoidal standing wave oscillator*, Fig. 12.4(a): $V(x,t) = A\cos\beta x \cos\omega_0 t$ and $I'(x,t) = A\sin\beta x \sin\omega_0 t$ in (12.21) yields

$$D_{\text{sin}} = \frac{2kTZ_0 G_a' \omega_0^2}{\pi^2 A^2}. \tag{12.22}$$

2. *Reflection pulse (linear and soliton) oscillators*, Figs. 12.4(b)(c): At the active circuit end of the line, an incident pulse is superposed with a reflected pulse. The joint voltage amplitude is about twice the incident pulse amplitude, or $V(0,t) \approx 2A\text{sech}^2(Bvt)$. In the soliton oscillator case, $V(0,t)$ may slightly differ from the linear superposition, due to the nonlinear interaction between the oncoming and reflected pulses, but this does not change the result significantly. Using (12.13) in (12.21), we obtain

$$D_{\text{pulse}} \approx \frac{kTZ_0 G_a' \omega_0^2 W}{3.76 A^2 d}. \tag{12.23}$$

Comparing (12.22) with (12.23), we find, in the case of the lumped noise perturbation,

$$\frac{D_{\text{pulse}}}{D_{\text{sin}}} \approx 1.3\frac{W}{d}, \tag{12.24}$$

if the standing wave and pulsed oscillators have the same amplitude for the same lumped noise level, and

$$\frac{D_{\text{pulse}}}{D_{\text{sin}}} \approx \frac{4W^2}{d^2}, \tag{12.25}$$

if they have the same power dissipation for the same lumped noise level. These results are exactly the same as in the case of distributed noise perturbation: (12.15) and (12.16). The results again indicate that a sharper pulse experiences the lumped noise at $x = 0$ for a shorter period of time, thus yielding a slower phase diffusion.

3. *Ring soliton oscillator*, Fig. 12.4(d):
   The voltage amplitude at the end points of the NLTL is given by $V(0,t) = A\text{sech}^2(Bvt)$, without soliton superposition. Using this and (12.13) in (12.21), we obtain:

$$D_{\text{soliton}} \approx \frac{kTZ_0G'_a\omega_0^2W}{7.5A^2d}. \tag{12.26}$$

   Based on (12.26) and (12.23), it appears that the ring soliton oscillator has better phase noise than the reflection soliton oscillator, if only the lumped noise sources are considered. However, the additional lumped noise generated by the termination of the ring soliton oscillator may have significant contribution to the phase noise.

So far we have considered only the direct phase perturbations along the tangential directions of the limit cycle, while not addressing the effect of the amplitude perturbation. As shown by Kaertner, the amplitude perturbation can contribute substantially to phase diffusion through amplitude-to-phase noise conversion in certain oscillators [78]. This effect is not of great importance in the standing wave and linear pulse oscillators, for the oscillation frequency of the linear transmission line with reasonably high Q is essentially independent of oscillation amplitude. In contrast, in the soliton pulse oscillator, the amplitude perturbation will translate to timing perturbations through soliton's amplitude-dependent propagation speed in the nonlinear transmission line, which can contribute significantly to phase noise. We now turn to this issue for the soliton oscillator.

12.3 Amplitude-to-Phase Noise Conversion

In this section, we examine how amplitude perturbation contributes to phase noise in soliton oscillators through amplitude-to-phase error conversion. We again use Kaertner's orthogonal projection scheme, but we do not carry it out in full to determine the exact amplitude-to-phase error conversion, which requires diagonalizing large matrices corresponding to oscillator's dynamical equations linearized in proximity of its limit cycle. Instead, we present a physically intuitive approach that captures the essence of amplitude-to-phase error conversion in soliton oscillators.

As discussed in Chapter 2, soliton pulse propagating in an NLTL can be described as $V = A\text{sech}^2[B(A)(x - v(A)t)] \approx I'$, where both

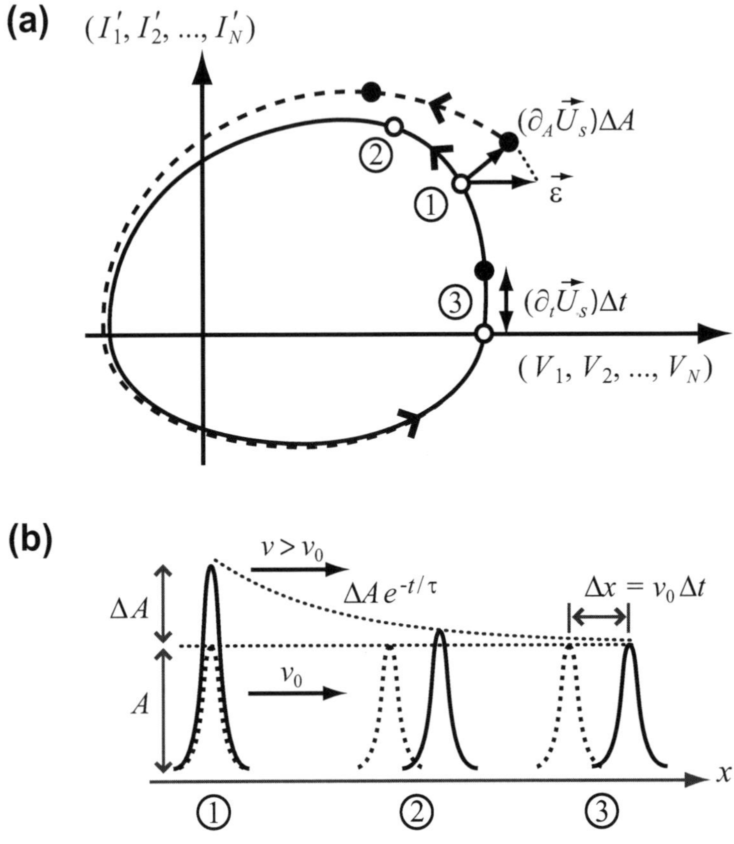

FIGURE 12.7
(a) Amplitude perturbation in the $2N$-D state space; (b) amplitude-to-phase error conversion via soliton's amplitude-dependent propagation speed.

the spatial pulse width $W \approx 1.763/B$ and wave velocity v depend on the amplitude A. In general, a taller soliton has a narrower width and propagates faster. The exact dependency is determined by the specific form of the line's nonlinearity or the capacitance-voltage relation of the varactors. With the pulse waveform above, the limit cycle $\vec{U}_s$ is now parameterized not only in terms of time t, but also in terms of amplitude A.[5] Just like $\partial_t \vec{U}_s$ giving the direction of phase/timing perturbation, $\partial_A \vec{U}_s$ defines the direction of amplitude perturbation in the state space. Following a procedure similar to that in Sec. 12.2, we now calculate the amplitude error by projecting a noise perturbation vector along the direction defined by $\partial_A \vec{U}_s$ [Fig. 12.7(a)].

12.3.1 Distributed Noise Sources

For distributed noise, $\vec{\varepsilon} = (\{\varepsilon_{G,n}\}, \{\varepsilon_{R,n}\})/\sqrt{LC}$ in Sec. 12.2.1 acts again as the noise perturbation vector. It yields the following amplitude error:

$$\Delta A = \frac{\vec{\varepsilon} \cdot \partial_A \vec{U}_s}{\left|\partial_A \vec{U}_s\right|^2} = \frac{\sum_n \varepsilon_{G,n}(\partial_A V_n) + \sum_n \varepsilon_{R,n}(\partial_A I'_n)}{\sqrt{LC}[\sum_n (\partial_A V_n)^2 + \sum_n (\partial_A I'_n)^2]}. \tag{12.27}$$

ΔA is Gaussian with zero mean and variance

$$\langle(\Delta A)^2\rangle = \frac{2kTZ_0}{LC} \frac{\sum_n G'(\partial_A V_n)^2 + R'(\partial_A I'_n)^2}{[\sum_n (\partial_A V_n)^2 + (\partial_A I'_n)^2]^2} dt. \tag{12.28}$$

Unlike phase perturbation, the amplitude perturbation puts the oscillation off the limit cycle and is corrected in finite time by the oscillator's dynamics. Thus, the amplitude error cannot accumulate indefinitely, but decays exponentially as $\Delta A e^{-t/\tau}$ (Fig. 12.7) where ΔA is the initial amplitude error and τ is the lifetime of the decay, which is determined by the oscillator's loss and gain characteristics. Since the soliton propagation velocity v is a function of A, this decaying amplitude perturbation will incur a velocity variation of Δv, according to

$$\Delta v = \frac{\partial v}{\partial A} \Delta A e^{-t/\tau},$$

which will translate to a total position shift of

$$\Delta x = \int_0^\infty \Delta v dt = \int_0^\infty \frac{\partial v}{\partial A} \Delta A e^{-t/\tau} dt = \frac{\partial v}{\partial A} \Delta A \tau,$$

[5]A similar parametrization of the limit cycle was used by H. A. Haus [84] in his study of the noise processes in soliton lasers.

as illustrated in Fig. 12.7(b). This is equivalent to a timing error of

$$\Delta t = \frac{\Delta x}{v} = \frac{\tau}{v}\frac{\partial v}{\partial A}\Delta A.$$

Therefore, the total contribution to the timing uncertainty by ΔA generated during $[t, t+dt]$ is

$$\langle(\Delta t)^2\rangle = \left(\frac{\partial v}{\partial A}\frac{\tau}{v}\right)^2 \langle(\Delta A)^2\rangle. \tag{12.29}$$

Plugging (12.28) into (12.29) and comparing the result with (12.1), we find the phase diffusion rate D due to the amplitude perturbation to be

$$D = \left(\frac{\partial v}{\partial A}\frac{\tau\omega_0}{v}\right)^2 \frac{kTZ_0}{LC}\left\langle \frac{\sum_n G'(\partial_A V_n)^2 + R'(\partial_A I_n')^2}{[\sum_n (\partial_A V_n)^2 + (\partial_A I_n')^2]^2}\right\rangle. \tag{12.30}$$

In continuous limit, this becomes

$$D = \left(\frac{\partial v}{\partial A}\tau\omega_0\right)^2 kTZ_0\left\langle \frac{\int g'(\partial_A V)^2 + r'(\partial_A I')^2 dx}{\left[\int (\partial_A V)^2 + (\partial_A I')^2 dx\right]^2}\right\rangle. \tag{12.31}$$

The phase diffusion rate depends on three factors: i) the sensitivity of amplitude to noise perturbations, captured by $\int(\partial_A V_n)^2 dx$ and $\int(\partial_A I_n')^2 dx$; ii) the sensitivity of velocity to amplitude $(\partial v/\partial A)$; iii) the lifetime τ of amplitude error, i.e., how fast the amplitude error is corrected by the oscillator's dynamics.

12.3.2 Lumped Noise Sources

Similar results can be obtained for the lumped noise sources described in Sec. 12.2.2:

$$D = \left(\frac{\partial v}{\partial A}\frac{\tau\omega_0}{v}\right)^2 \frac{kTZ_0}{LC}\left\langle \frac{G_a'(t)[\partial_A V_1(t)]^2}{[\sum_n (\partial_A V_n)^2 + (\partial_A I_n')^2]^2}\right\rangle \tag{12.32}$$

for an NLTL, and

$$D = \left(\frac{\partial v}{\partial A}\tau\omega_0\right)^2 kTZ_0\left\langle \frac{G_a'(t)[\partial_A V(0,t)]^2}{\left[\int (\partial_A V)^2 + (\partial_A I')^2 dx\right]^2}\right\rangle \tag{12.33}$$

in the continuous limit.

12.3.3 Indirect vs. Direct Phase Perturbations

The total phase diffusion rate in a soliton oscillator is a sum of D_t caused by direct phase perturbations and D_A induced by amplitude perturbations. This simple addition is possible because the amplitude perturbation [e.g., (12.27)] and the phase perturbation [e.g., (12.6)] are orthogonal, or uncorrelated to each other. This orthogonality can be checked by noticing that $\partial_A V$ and $\partial_A I'$ in (12.27) are even functions of x, whereas $\partial_t V$ and $\partial_t I'$ in (12.6) are odd.

Given this simple addition, we may directly compare D_A and D_t to see which process contributes more to the total phase diffusion rate. For a concrete comparison, we use the soliton waveform of $V = I' = A\text{sech}^2[B(A)(x - v(A)t)]$ with specific expressions for $B(A)$ and $v(A)$ found in Chapter 2. With this waveform, (12.31) yields

$$D_A = \left(\frac{\partial v}{\partial A}\right)^2 kTZ_0\omega_0^2\tau^2\frac{r' + g'}{3.8W}. \tag{12.34}$$

for distributed noise sources in both reflection and ring soliton oscillators. Comparing this with D_t of (12.14) and (12.17) caused by the same distributed noise sources, we have

$$\frac{D_A}{D_t} \approx 2\left(\frac{\partial v}{\partial A}\frac{A}{W}\tau\right)^2, \tag{12.35}$$

Exactly the same relationship holds in the case of lumped noise perturbations, according to (12.21) and (12.33), if $G'_a(t)$ is a constant. As seen, the ratio depends largely on the velocity-amplitude relation and the lifetime of amplitude errors. D_A can be far larger than D_t.

12.4 Experimental Verification

12.4.1 Oscillator Prototypes

For experimental proof of concept, we design the sinusoidal standing wave oscillator, reflection linear pulse oscillator, and reflection soliton oscillators of Fig. 12.4(a)~(c) in the lower microwave region, and implement them on printed circuit boards using discrete components. The reflection soliton oscillator is adopted from [67], whose circuit and detailed operation can be found in Chapter 10. The same circuit board layout and circuit topology are used to construct the standing wave and linear pulse oscillators, whose transmission lines are implemented using

TABLE 12.1
Waveform parameters of three distributed oscillators.

Oscillator type	Sinusoidal standing	Linear pulse	Soliton pulse
Oscillation frequency (MHz)	95.9	93.3	82.4
Maximum frequency content (MHz)	95.9	373.2	1071.2
Amplitude (volt)	2.0	2.2	1.8
Pulse Width (section)	-	9.5	3.0

linear LC components. In addition, the component parameters of the active circuit are adjusted to produce sinusoidal standing wave or pulse oscillations.

The oscillators produced the desired oscillation waveforms. A 6-V battery was used for stable power supply. The waveforms were monitored using an Agilent Infiniium 54855A 6-GHz oscilloscope with Agilent 1156A 100 kΩ active probes. Fig. 12.8 shows steady-state oscillation waveforms measured at different line positions in each of the three oscillators. Waveform parameters are summarized in Table 12.1. In the reflection soliton oscillator, the soliton pulse undergoes soliton-shaping and damping processes as it propagates along the nonlinear transmission line with amplitude A ranging from 1.6 to 2.0 V. An average value of 1.8 V is used here for simplicity of analysis. The oscillation frequencies are around 100 MHz for all oscillators. The harmonic contents of the pulse oscillators extend well into the microwave regime[6].

12.4.2 Phase Noise Measurement

Fig. 12.9 shows the phase noise measurement setup. The transmission line's open end is connected to an Agilent E5052B signal source analyzer, which has built-in phase noise measurement function, via an L-section impedance matching network and a Mini-Circuits ZFL-11AD amplifier. The matching network is to ensure that the line sees a high impedance at what is designed to be an open end, so that almost total reflection can occur as intended. The amplifier enhances the weak signal coupled out from the oscillator, and also isolates the oscillators from the signal source analyzer so that the internal switching inside the analyzer will not

[6]The cutoff is set as the highest harmonics whose power is within -20 dB of the power at the fundamental frequency.

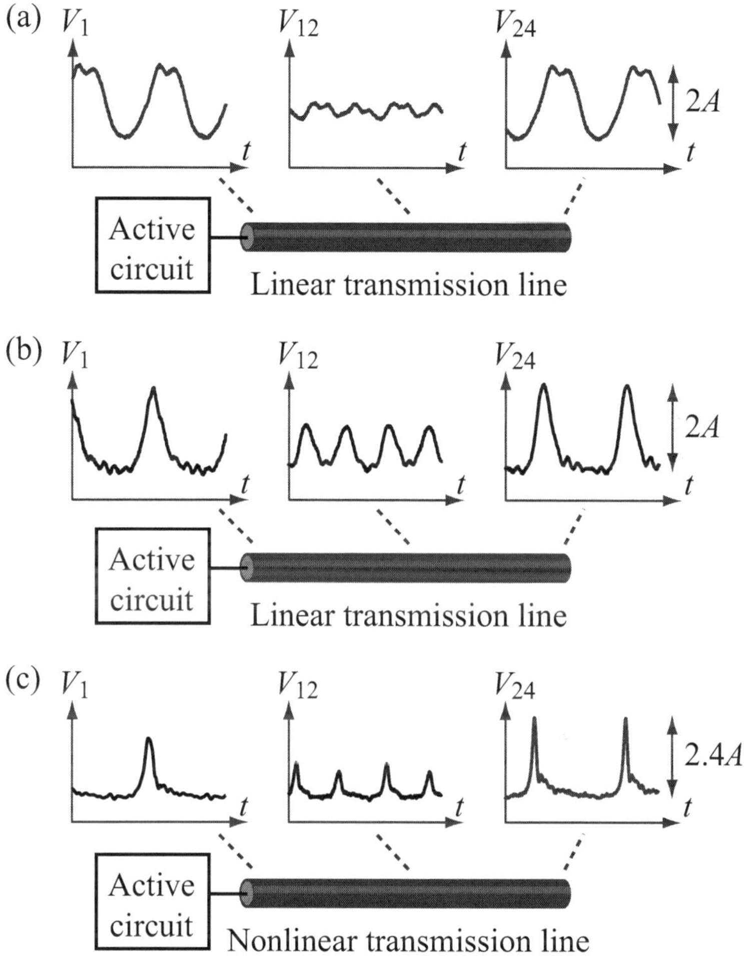

FIGURE 12.8
(a) Near-sinusoidal standing wave oscillator; (b) Reflection linear pulse oscillator; (c) Reflection soliton oscillator.

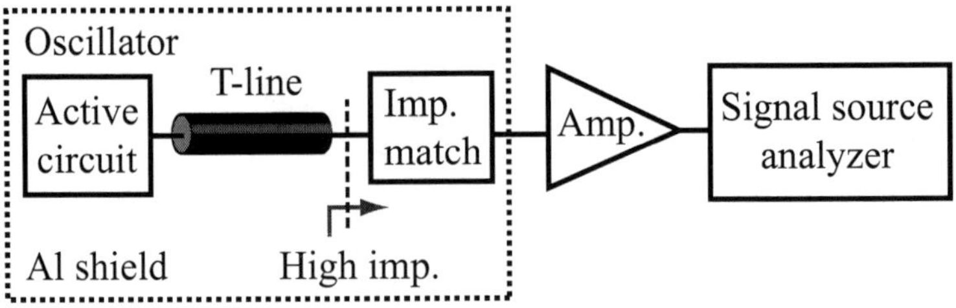

FIGURE 12.9
Phase noise measurement setup.

perturb the oscillator's operation. In addition, the oscillator, battery, and matching network are enclosed in an aluminum shield to minimize external interference.

Fig. 12.10 shows the measured phase noise of the three oscillators. The phase noise largely follows an f^{-2} trend (with a slight f^{-3} tendency at low offset frequencies), indicating that the noise sources in all oscillators are dominantly white. The measurements agree with the core of our analysis in the previous section: the linear pulse oscillator has a better phase noise than the standing wave oscillator, as predicted. The soliton oscillator, however, has the worst phase noise despite the soliton's sharpness. This is due to its amplitude-to-phase noise conversion, as we will show in the following subsections.

12.4.3 Intensity of Noise Sources

To quantitatively compare the measured phase noise with our calculation, we first determine the intensity of the noise sources in the oscillators.

We start with the lumped noise, which is generated from the active circuit in the three oscillators. We observed in all three oscillators that the active circuit remains inactive except at the rising edge of the voltage signal $V_1(t)$ at the left end of the line. (In continuous coordinate, $V_1(t)$ corresponds to $V(0,t)$). The rising edge of V_1 triggers transistors N_1, P_1 and P_2 in Fig. 10 so that each pulls a constant current of about 10 mA until V_1 reaches maximum. During this process energy is injected to provide gain, and also, appreciable noise is injected, as illustrated in Fig. 12.11.

The intensity of the active noise could be measured at *dc* by biasing the circuit with the same currents (10 mA) as experienced at the rising edge. Such measurement, however, was stymied in our experiment, for

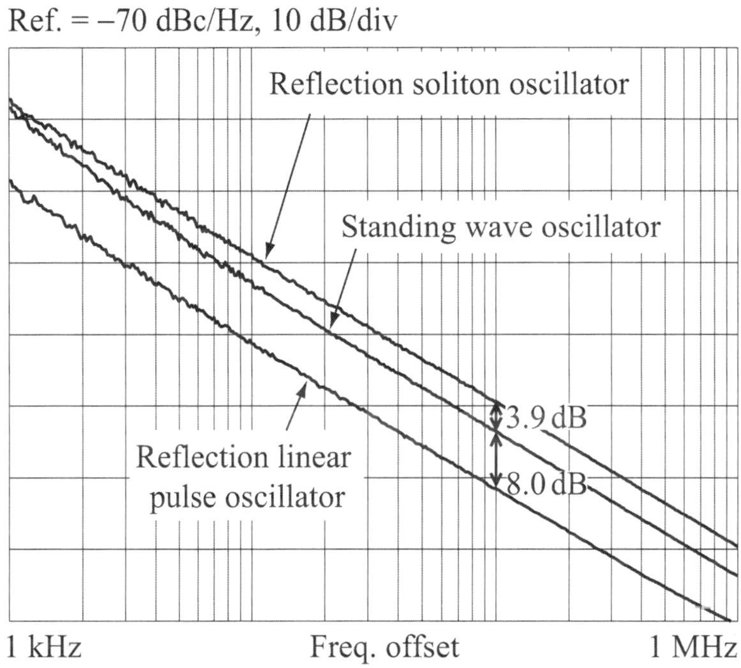

FIGURE 12.10
Measured phase noise for the three oscillators.

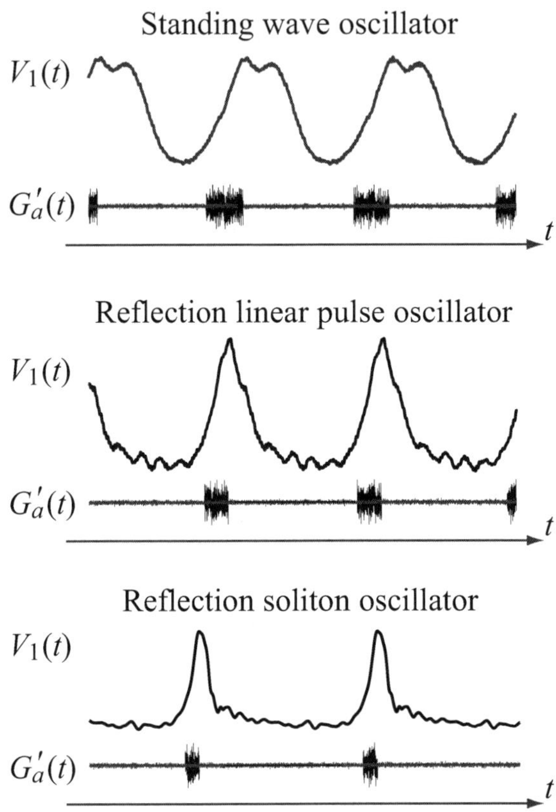

FIGURE 12.11
Time-domain illustration of the lumped noise from the active circuits.

10 mA *dc* current exceeds the *dc* breakdown current of transistor N_1. Even if transistors that could survive the *dc* current are chosen, the transistors' temperature with the large, continued *dc* current would be higher than its actual value during the oscillator operation, thus, not necessarily providing a faithful replica of the actual active noise level.

We estimate the intensity of the active noise using SPICE simulation, in which the circuit is bias at *dc* with the same currents (10 mA) as experienced at the rising edge, while the voltage noise (intensity of i_a in Fig. 12.6 multiplied by Z_0^2) intensity is measured at the output of the active circuit with a 50 Ω load. Using this method, the intensity of the active circuit noise during the rising edge of V_1 was estimated to be $4kTZ_0G'_a(t) = 8.5 \times 10^{-17}$ V^2/Hz for all three oscillators. The active noise was negligibly smaller during the rest of V_1 dynamics (Fig. 12.11).

For the phase noise calculation, we can effectively reduce this time-varying active noise to a constant noise. Imagine we use $4kTZ_0G'_a(t) = 8.5 \times 10^{-17}$ V^2/Hz throughout the entire V_1 dynamics in calculating the phase diffusion rate [(12.20)/(12.21)], as if it persisted all the time. During the "flat" part of the V_1 waveform (i.e., during the absence of pulse events or transition edges), the fictitious noise does not contribute to (12.20)/(12.21), as $\dot{V}_1(t) = 0$; thus, its use for the flat duration is justified. During the falling edges, the fictitious noise contributes the same amount to (12.20)/(12.21) as the real noise during the rising edges does, for the oscillation waveforms are almost symmetric about their maxima. Hence, if we use half of the actual noise intensity injected only during the rising edge, that is, $4kTZ_0G'_a = 4.2 \times 10^{-17}$ V^2/Hz, as a constant noise intensity the entire time, it will yield the correct phase diffusion rate. We will use this constant noise in the following analysis for all three oscillators.

The distributed noise generated from loss elements in the transmission line is far smaller in our experimental circuits. For both linear and nonlinear lines, the measured quality factor Q is about 100 at the oscillation frequency and the measured Z_0 is 50 Ω. Thus, the total noise power of the lossy lines is $4kTZ_0(\pi/Q) = 2.6 \times 10^{-20}$ V^2/Hz, which is 3 orders of magnitude smaller than the lumped active noise. In the following calculation of phase noise, we use only the dominant lumped active noise.

12.4.4 Measurement–Theory Comparison

By plugging the relevant waveform parameters of Table 12.1 and the effective constant active noise level into (12.22) and (12.23), we find the phase diffusion rate, D_t, due to direct tangential phase perturbation, for all three oscillators. See Table 12.2. The standing wave, reflection linear

Oscillator type	Sinusoidal standing	Linear pulse	Soliton pulse
D_t (direct phase perturbation)(Hz)	0.19	0.079	0.029
D_A (amplitude-phase conversion)(Hz)	-	-	0.33
$D_{total} = D_t + D_A$ (Hz)	0.19	0.079	0.36
Phase noise, 100kHz offset (dBc/Hz)	−120.2	−124.0	−117.4

TABLE 12.2
Calculated phase diffusion rates and phase noise.

pulse, and reflection soliton oscillators line up in the decreasing order of D_t, owing to the decreasing pulse width as explained earlier. In the soliton oscillator, although D_t is substantially small, we must also consider the phase diffusion rate D_A due to the indirect phase perturbation caused by amplitude perturbation and its conversion to phase error. It is shown in Table 12.2 as well.

The calculation of D_A in the soliton oscillator is done as follows. Eq. (12.35) may be rewritten into

$$\frac{D_{A,\text{ soliton}}}{D_{t,\text{ soliton}}} \approx 2\left(\frac{\tau A}{T^2}\frac{\partial T}{\partial A}\frac{2d}{W}\right)^2, \tag{12.36}$$

by using soliton's round-trip time $T = 2d/v$ in the nonlinear line. To estimate $\partial T/\partial A$ and τ, we perturb the soliton's amplitude with a short pulse produced by an Agilent 81150A function generator [Fig. 12.12(a)]. From the decaying behavior of the amplitude perturbation, we measure A and T of the decaying pulse as illustrated in Fig. 12.12(b). Then, from the slopes of $\log|\Delta A|$ vs. t and T vs. A [Fig. 12.12(c)(d)], we obtain $\tau = (1.50 \pm 0.05)T$ and $A\partial T/\partial A = -1.20 \pm 0.15$ ns. (The same scheme can be implemented in simulation to estimate these parameters in designing the oscillators.) Plugging these measured values in (12.36) yields $D_{A,\text{soliton}} = (11.3 \pm 3.5)D_{t,\text{ soliton}} = 0.33 \pm 0.10$ Hz, which enter Table 12.2. In the soliton oscillator, the indirect phase perturbation contributes about 10 times more to the total phase noise than the direct phase perturbation. The corresponding calculated phase noise values for all oscillators at 100 kHz offset are also shown in Table 12.2. The measurement errors in τ and $A\partial T/\partial A$ give a ±1.2 dB variation in the phase noise calculation for soliton pulse oscillator.

Since the active noise level was only estimated in simulation (Subsec. 12.4.3), it is most meaningful to compare the relative differences of

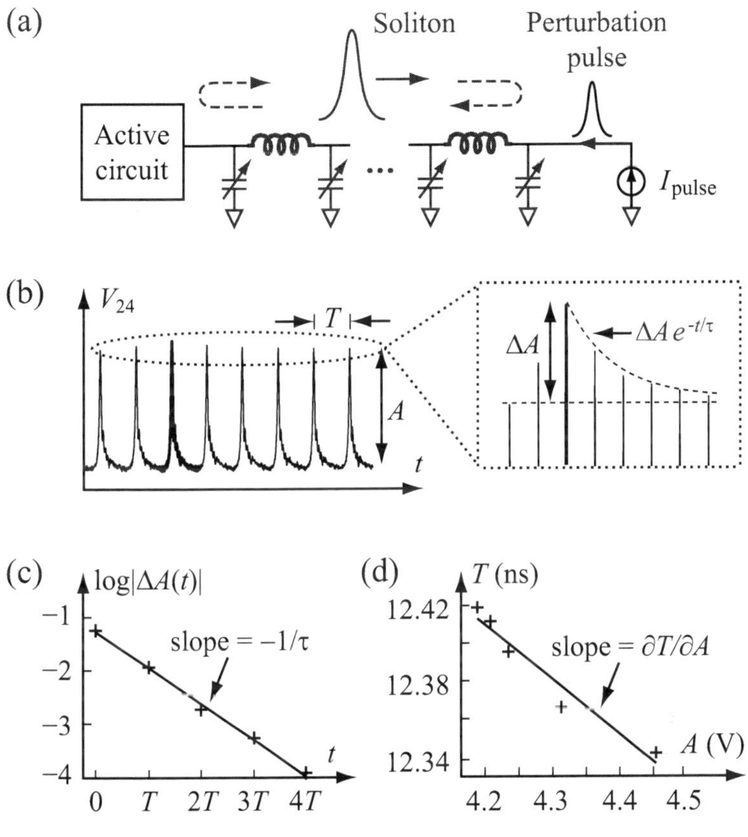

FIGURE 12.12
Measured decay behavior of amplitude perturbation in the soliton oscillator. (a) Measurement setup. (b) Decay behavior after an amplitude perturbation ΔA. (c) τ extraction. (d) $\partial T/\partial A$ extraction.

the measured and calculated phase noise amongst the oscillators. (The actual active noise level, albeit not exactly known, is expectedly the same for all three oscillators).

First, when compared to the standing wave oscillator, the soliton oscillator has a 2.8 dB worse phase noise in calculation (Table 12.2), which is close to the 3.9 dB worse phase noise in measurement (Fig. 12.10). These results show that in the soliton oscillator, despite the soliton's sharpness that would yield superb phase noise if only the direct phase perturbation existed, the indirect phase perturbation through amplitude-to-phase error conversion significantly contributes to phase noise, offsetting the benefit of the soliton's sharpness.

Second, in comparison to the standing wave oscillator, the linear pulse oscillator has a 3.8 dB better phase noise in calculation (Table 12.2) and a 8.0 dB better phase noise in measurements (Fig. 12.10). This shows the phase noise superiority of pulse oscillators in absence of amplitude-to-phase noise conversion[7].

12.5 Summary

In this chapter, we examine how noise perturbs the collective oscillation of a distributed oscillator and we provide a physically-intuitive analysis that identifies two major contributions to the phase noise of soliton oscillators: direct phase perturbation and amplitude-to-phase noise conversion through soliton's amplitude-dependent velocity. If only the former existed, the soliton oscillator would show superb phase noise performance, thanks to the sharpness of soliton pulses. In our prototypes, however, this benefit is significantly offset by the indirect phase perturbation through amplitude-to-phase error conversion. Finally, the analysis provides a couple of useful design strategies to mitigate the amplitude-to-phase noise conversion, which largely depends on the lifetime τ of amplitude error and the sensitivity of oscillation period T to amplitude A. Both can be reduced by a proper active circuit design that quickly corrects amplitude error and introduces amplitude-dependent delay to compensate for the variation in T caused by amplitude errors.

[7]The numerical difference may be attributed to that in the actual linear pulse oscillator; the injected noise could have been smaller due to the detailed difference in its operation.

Bibliography

[1] A. C. Scott, F. Y. F. Chu, and D. W. McLaughlin, "The soliton: a new concept in applied science," *Proc. IEEE*, vol. 61, pp. 1443–1483, Oct. 1973.

[2] N. J. Zabusky and M. D. Kruskal, "Interactions of solitons in a collisionless plasma and the recurrence of initial states," *Phys. Rev. Lett.*, vol. 15, no. 6, pp. 240–243, 1965.

[3] P. G. Drazin and R. S. Johnson, *Solitons: an introduction.* NY: Cambridge University Press, 1989.

[4] A. T. Filippov, *The versatile soliton.* Boston: Birkä user, 2000.

[5] M. Remoissenet, *Waves Called Solitons: Concepts and Experiments.* New York, NY: Springer, 1999.

[6] J. S. Russel, "Report on waves," *Fourth meeting of the British Asoication for the Advancement of Science, York. UK*, pp. 311–3902, Sept. 1844.

[7] E. Fermi, J. Pasta, and S. Ulam, "Studies of nonlinear problems. i," *Los Alamos Sci. Lab. Rep.*, vol. LA-1940, 1955.

[8] A. C. Newell, *Nonlinear Wave Motion.* Providence, RI: American Mathematical Society, 1974.

[9] A. Hasegawa and F. Tappert, "Transmission of stationary nonlinear optical pulses in dispersive dielectric fibres. i. anomalous dispersion," *Applied Phy. Lett.*, vol. 23, pp. 142–144, Aug. 1973.

[10] R. Hirota and K. Suzuki, "Theoretical and experimental studies of lattice solitons in nonlinear lumped networks," *Proc. IEEE*, vol. 61, pp. 1483–1491, Oct. 1973.

[11] M. J. Rodwell, S. T. Allen, R. Y. Yu, M. G. Case, U. Bhattacharya, M. Reddy, E. Carman, M. Kamegawa, Y. Konishi, J. Pusl, and R. Pullela, "Active and nonlinear wave propagation devices in ultrafast electronics and optoelectronics," *Proc. IEEE*, vol. 82, pp. 1037–1059, July 1994.

[12] D. Salameh and D. Linton, "Microstrip gaas nonlinear transmission-line (nltl) harmonic and pulse generators," *IEEE Trans. Microw. Theory Tech.*, vol. 47, pp. 1118–1122, July 1999.

[13] M. Kahrs, "50 years of rf and microwave sampling," *IEEE Trans. Microw. Theory Tech.*, vol. 51, pp. 1787–1805, June 2003.

[14] R. Y. Yu, J. Puls, K. Yoshiyuki, M. Case, K. Masayuki, and M. Rodwell, "A time-domain millimeter-wave vector network analyzer," *IEEE Trans. Microw. Theory Tech.*, vol. 2, pp. 319–321, Aug. 1992.

[15] B. Elliot, "High-sensitivity picosecond time-domain reflectometry," *IEEE Trans. Instrum. Meas.*, vol. IM-25, Dec. 1976.

[16] D. S. Ricketts, X. Li, and D. Ham, "Electrical soliton oscillator," *IEEE Trans. Microw. Theory Tech.*, vol. 54, pp. 373–382, Jan. 2006.

[17] A. Singer, A. Oppenheim, and G. Wornell, "Detection and estimation of multiplexed soliton signals," *IEEE Trans. Signal Process.*, vol. 47, pp. 2768 –2782, Oct. 1999.

[18] D. Ham, X. Li, S. A. Denenberg, T. H. Lee, and D. S. Ricketts, "Ordered and chaotic electrical solitons: communication perspectives," *IEEE Commun. Mag.*, vol. 44, no. 12, pp. 126–135, 2006.

[19] A. C. Scott, *Nonlinear Science: Emergence and Dynamics of Coherent Structures.* Oxford: Oxford Univ. Press, 2003.

[20] D. J. Korteweg and G. de Vries, "On the change of form of long waves advancing in a rectangular canal, and on a new type of long stationary waves," *Phil. Mag.*, vol. 39, pp. 422–443, 1895.

[21] D. Jäger, "Experiments on KdV solitons," *J. Phys. Soc. Jpn.*, vol. 51, pp. 1686–1693, May 1982.

[22] C. C. Cutler, "The regenerative pulse generator," *Proc. IRE*, vol. 43, pp. 140–148, Feb. 1955.

[23] R. Landauer, "Parametric amplification along nonlinear transmission lines," *J. Appl. Phys.*, vol. 31, pp. 479–484, Mar. 1960.

[24] A. J. DeMaria, D. A. Stetser, and H. Heynau, "Self mode-locking of lasers with saturable absorbers," *Appl. Phys. Lett.*, vol. 8, pp. 174–176, Apr. 1966.

[25] K. Tamura, H. A. Haus, and E. P. Ippen, "Self-starting additive pulse mode-locked erbium fibre ring laser," *Electron. Lett.*, vol. 28, pp. 2226–2228, Nov. 1992.

[26] M. Toda, "Vibration of a chain with nonlinear interaction," *J. Phys. Soc. Jpn.*, vol. 22, pp. 431–436, Feb. 1967.

[27] M. Toda, "Wave propagation in anharmonic lattices," *J. Phys. Soc. Jpn.*, vol. 23, pp. 501–506, Sept. 1967.

[28] C. S. Gardner, J. M. Greene, M. D. Kruskal, and R. M. Miura, "Method for solving the Korteweg-deVries equations," *Phys. Rev. Lett.*, vol. 19, pp. 1095–1097, Nov. 1967.

[29] P. Lax, "Integrals of nonlinear equations of evolution and solitary waves," *Comm. Pure and Appl. Math.*, vol. 21, pp. 467–490, 1968.

[30] D. Jäger, "Characteristics of travelling waves along the non-linear transmission lines for monolithic integrated circuits: a review," *Ing. J. Electronics*, vol. 58, pp. 649–669, 1985.

[31] L. F. Mollenauer, R. H. Stolen, and J. P. Gordon, "Experimental observation of picosecond pulse narowing and solitons in optical fibers," *Phys. Rev. Lett.*, vol. 45, pp. 1095–1098, 1980.

[32] M. Tan, C. Y. Su, and W. J. Anklam, "7 x electrical pulse compression on an inhomogenous nonlinear transmission line," *Electron. Lett.*, vol. 24, pp. 213–215, Feb. 1988.

[33] M. G. Case, *Nonlinear Transmission Lines for Picosecond Pulse, Impulse and Millimeter-Wave Harmonic Generation.* PhD thesis, University of California Santa Barbara, 1993.

[34] D. W. van der Weide, "Delta-doped schottky diode nonlinear transmission lines for 480-fs, 3.5-v transients," *Applied Phys. Lett.*, vol. 65, pp. 881–883, Aug. 1994.

[35] A. T. Allen, *Schottky Diode Integrated Circuits for Sub-Millimeter-Wave Applications.* PhD thesis, University of California Santa Barbara, 1994.

[36] G. J. Ballantyne, P. T. Gough, and D. P. Taylor, "Periodic solutions of toda lattice in loop nonlinear transmission line," *Elec. Lett.*, vol. 29, pp. 607–609, Apr. 1993.

[37] D. S. Ricketts, X. Li, M. DePetro, and D. Ham, "A self-sustained electrical soliton oscillator," *IEEE MTT-S Int. Microwave Symp. Dig.*, June 2005.

[38] D. S. Ricketts and D. Ham, "A chip-scale electrical soliotn modelocked oscillator," *IEEE Int. Solid-State Circuits Conf.*, p. 432, 2006.

[39] E. Afshari and A. Hajimiri, "Nonlinear transmission lines for pulse shaping in silicon," *IEEE J. Solid-State Circuits*, vol. 40, pp. 744 – 752, March 2005.

[40] E. Afshari, H. Bhat, X. Li, and A. Hajimiri, "Electrical funnel: a new signal combining method," *International Solid-State Circuits Conference*, pp. 240–025, Feb. 2006.

[41] R. Hirota, *The Direct Method in Soliton Theory.* Cambridge, UK: Cambridge University Press, 2004.

[42] E. Ott and R. N. Sudan, "Damping of solitary waves," *The Physics of Fluid*, vol. 13, pp. 1432–1434, June 1970.

[43] A. G. Greenhill, *The Applications of Elliptic Functions.* NY: MacMillan and Co., 1892.

[44] G. L. Baker and J. A. Blackburn, *The Pendulum: A Case Study in Physics.* USA: Oxford University Press, 2005.

[45] H. A. Haus, "Mode-locking of lasers," *IEEE J. Sel. Top. Quant. Elect.*, vol. 6, pp. 1173–1185, Nov. 2000.

[46] F. Calogero and A. Degasperis, *Spectral Transform and Solitons: Tools to Solve and Investigate Nonlinear Evolution Equations.* NY: North Holland, 1982. p. 49.

[47] I. M. Gel'fand and B. M. Levitan, "On the determination of a differential equation from its spectral function," *Am. Math. Transl.*, vol. 1, pp. 253–304, 1951.

[48] G. B. Whitham, *Linear and Nonlinear Waves.* NY: Wiley-Interscience, 1974.

[49] D. J. Griffiths, *Introduction to Quantum Mechanics.* NJ: Prentice Hall, 1994.

[50] R. Hirota, "Exact solution of the Korteweg-de Vries equation for multiple collisions of solitons," *Phys. Rev. Lett.*, vol. 27, pp. 1192–1194, Nov. 1971.

[51] R. M. Miura, C. S. Gardner, and M. Kruskal, "Korteweg-de Vries equation and generalizations. ii. existence of conservation laws and constants of motion," *J. Math. Phys.*, vol. 9, pp. 1204–1209, 1968.

[52] T. Taniuti and C. C. Wei, "Reductive perturbation method in nonlinear wave propagation," *J. Phys. Soc. Jpn.*, vol. 21, pp. 209–212, Jan. 1968.

[53] H. Nagashima and Y. Amagishi, "Experiment on the toda lattice using nonlinear transmission lines," *J. Phys. Soc. Jpn.*, vol. 45, pp. 680–688, Aug. 1978.

[54] H. Leblond, "The reductive perturbation method and some of its applications," *J. Phys. B: At. Mol. Opt. Phys.*, vol. 41, p. 043001, Feb. 2008.

[55] M. Toda, *Nonlinear Waves and Solitons*. Dordrecht: Kluwer, 1989.

[56] R. Y. Yu, M. Reddy, J. Pusl, S. T. Allen, M. Case, and M. J. W. Rodwell, "Millimeter-wave on-wafer waveform and network measurements using active probes," *IEEE Trans. Microw. Theory Tech.*, vol. 43, pp. 721–729, Apr. 1995.

[57] M. J. W. Rodwell, M. Karnegawa, M. C. R. Yu, E. Carman, and K. S. Giboney, "Gaas nonlinear transmission lines for picosecond pulse generation and millimeterwave sampling," *IEEE Trans. Microw. Theory Tech.*, vol. 39, pp. 1194–1204, July 1991.

[58] M. G. Case, *Nonlinear Transmission Lines for Picosecond Pulse, Impulse and Millimeter-Wave Harmonic Generation*. PhD thesis, University of California Santa Barbara, 1993.

[59] B. Wedding and D. Jäger, "Phase-matched second harmonic generation and parametric mixing on nonlinear transmission lines," *Electron. Lett.*, vol. 17, pp. 76–77, 1981.

[60] W. Andress and D. Ham, "Standing wave oscillators utilizing wave-adaptive tapered transmission lines," *IEEE J. Solid-State Circuits*, vol. 40, Mar. 2005.

[61] D. Ham and A. Hajimiri, "Concepts and methods in optimization of integrated *lc* vcos," *IEEE J. Solid-State Circuits*, vol. 36, pp. 896–909, June 2001.

[62] G. J. Ballantyne, P. T. Gough, and D. Taylor, "A baseband soliton oscillator," *Chaos, Solitons and Fractals*, vol. 5, pp. 1013–1029, June 1995.

[63] G. J. Ballantyne, *Periodically Amplified Soliton Systemsn*. PhD dissertaion, University of Canterbury, New Zealand, 1994.

[64] L. A. Glasser and H. A. Haus, "Microwave mode locking at x-band using solid-state devices," *IEEE Trans. Microw. Theory Tech.*, vol. 26, pp. 62–69, Feb. 1978.

[65] S. Nogi and K. Fukui, "A method of reducing reflection for a soliton at the termination of a nonlinear lc line.," *Trans. IECE of Jpn.*, vol. 64, pp. 437–438, June 1982.

[66] S.-J. Kim, Y.-G. Lee, S.-K. Yun, and H.-Y. Lee, "Realization of high Q inductors using wirebonding technology," *IEEE AP-ASIC*, pp. 13–16, 1999.

[67] O. O. Yildirim, D. Ricketts, and D. Ham, "Reflection soliton oscillator," *IEEE Trans. Microw. Theory Tech.*, vol. 57, pp. 2344–2353, Oct. 2009.

[68] O. O. Yildirim, D. Ricketts, and D. Ham, "Reflection soliton oscillator," *IEEE Trans. Microw. Theory Tech.*, vol. 57, pp. 2344–2353, Oct. 2009.

[69] A. A. Tsonis, *Chaos: From theory to applications.* New York: Plenum, 1992.

[70] T. Stojanovski, J. Pihl, and L. Kocarev, "Chaos-based random number generators. part ii: practical realization," *IEEE Trans. Circuits Syst. -I*, vol. 48, pp. 382–385, Mar. 2001.

[71] L. M. Pecora and T. L. Carroll, "Synchronization in chaotic systems," *Phys. Rev. Lett.*, vol. 64, pp. 821–824, Feb. 1990.

[72] U. Parlitz, L. O. Chua, L. Kocarev, K. S. Halle, and A. Shang, "Transmission of digital signals by chaotic synchronization," *Int. J. Bifurcation Chaos*, vol. 2, pp. 973–977, Dec. 1992.

[73] K. M. Cuomo and A. V. Oppenheim, "Synchronization of Lorenz-based chaotic circuits with applications to communications," *IEEE Trans. Circuits Syst. -II*, vol. 40, pp. 626–633, Oct. 1993.

[74] C. E. Shannon, "Communication theory of secrecy systems," *Bell Syst. Technic. J.*, vol. 28, no. 4, pp. 656–715, 1949.

[75] C. Kurtsiefer, P. Zarda, M. Halder, H. Weinfurter, P. M. Gorman, P. R. Tapster, and J. G. Rarity, "Quantum cryptography: A step towards global key distribution," *Nature*, vol. 419, p. 450, Oct. 2002.

[76] M. J. W. Rodwell, M. Karnegawa, M. C. R. Yu, E. Carman, and K. S. Giboney, "Gaas nonlinear transmission lines for picosecond pulse generation and millimeterwave sampling," *IEEE Trans. Microw. Theory Tech.*, vol. 39, pp. 1194–1204, July 1991.

[77] M. Lax, "Classical noise. V. Noise in self-sustained oscillators," *Phys. Rev.*, vol. 160, pp. 290–307, Aug. 1967.

[78] F. X. Kaertner, "Determination of the correlation spectrum of oscillators with low noise," *IEEE Trans. Microw. Theory Tech.*, vol. 37, pp. 90–101, Jan. 1989.

[79] F. X. Kaertner, "Analysis of white and $f^{-\alpha}$ noise in oscillators," *Int. J. Circuit Theory Applicat.*, vol. 18, pp. 485–519, 1990.

[80] D. B. Leeson, "A simple model of feedback oscillator noise spectrum," *Proc. IEEE*, vol. 54, pp. 329–330, Feb. 1966.

[81] A. Hajimiri and T. H. Lee, "A general theory of phase noise in electrical oscillators," *IEEE J. Solid-State Circ.*, vol. 33, pp. 179–194, Feb. 1998.

[82] J. A. Gubner, *Probability and random processes for electrical and computer engineers.* NY: Cambridge University Press, 2006.

[83] M. Lax and W. H. Louisell, "Quantum noise IX: Quantum Fokker-Planck solution for laser noise," *IEEE J. Quantum Elect.*, vol. QE-3, pp. 47–58, Feb. 1967.

[84] H. A. Haus and Y. Lai, "Quantum theory of soliton squeezing: a linearized approach," *J. Opt. Soc. Am. B*, vol. 7, pp. 386–392, Mar. 1990.

Index